A RUGGED LIFE

50 Years of Fieldwork in the Canadian Wilderness

BY CHRIS CLEMENT

The author working in the southern Rocky Mountains on the High Elevation Grassland Project in 2020.
Photo by Christiane Brouwer.

First edition, first printing
Cover artwork by The Beacon Design Collective

This book is dedicated to Jim van Barneveld, who was my mentor and teacher for the first ten years of my vegetation ecologist career, as well as all the other government scientists that I had the privilege to work with and learn from in that period of my life.

"Look deep into nature, and then you will understand everything better."

– Albert Einstein

CONTENTS

AUTHOR BIO

Chris Clement started his career working in the Canadian wilderness as an inexperienced forest technology graduate in 1970.

Over the next fifty years of tramping through the endless, rugged mountains of BC, the vast boreal muskeg of Alberta, and the sprawling, windblown grasslands of Saskatchewan he became an expert on vegetation ecology and honed his skills at ecosystem and soil description. Along the way he has been stalked by bears, bushwhacked through countless kilometres of tangled brush, dodged behemoth logging trucks, and completed helicopter surveys in the wildest areas of western Canada. His story includes fascinating accounts of work done in Gwaii Haanas, northern BC, the west Kootenays, the totally pristine Charlotte Alplands, and the southern Rocky Mountains. His life has been one of dedication to mapping and understanding the native vegetation of a huge area of western Canada. His training started with the government of BC where he worked with a variety of esteemed scientists. Since that time, he has worked for small and large consulting companies, ran his own business for 12 years, and toiled independently. Now he is passing the torch to younger generations of scientists. He has travelled up and down the unexplored parts of the west coast, staying in rough logging camps, working in the seething industrial tar sands, conducting rare plant surveys and wetland assessments on the rolling prairies, and done research on the windswept Arctic tundra near Cambridge Bay. Now at the age of 72 this book describes many of his challenging adventures and the never-ending quest for demanding fieldwork. Join him as he takes you on an epic journey through the Canadian wilderness.

ACKNOWLEDGEMENTS

A huge thank you to my son Brian for all his encouragement, writing advice, editing, and most of all for providing the impetus for me to complete this book.

To my wife Anne, who has endured my endless forays into the wilderness, not knowing if I would ever return from the dangers I faced; thanks for allowing me to become intimate with Mother Nature and learn a few of her secrets.

My deep appreciation to Steve Lord, who was my partner in the summer of 71, shared the near-death experiences with me, watched my back when we were in the wilderness, and to this day is still a loyal and trusted friend.

A big thank you to all my field assistants over the years, including Steve Thompson, Paul Nobu Ono, Rick Ferster, Terry Wood, Craig Bissett, Tom Franklin, John Klarer, John Koji Ono, Mike Robinson, Peter Tatroff, Brent Keeping, Colin Clement, Mitch Williams, Rod Dalziel, Alec MacLennan, Shawn Munro, James Wadsworth, Casey Holmes Jr., Kiley Gibson, Teia Clement, Nadine Clifton, Jenn Lange, and Chloe Franklin. They were all excellent diggers of soil pits and put up with my insufferable thirst for vegetation data and love of nature.

A huge note of appreciation to the terrain mappers who always made my job easier, including Denny Maynard, Polly Uunila, Deepa

Spaeth, Michelle Burfitt, Ron Kowall, and Dr. David Huntley.

A debt of gratitude to Dr. Jim Pojar and Andy MacKinnon (and others) who produced the awesome plant identification books for BC.

Thanks also to all the helicopter pilots, who over the years flew me to and from my work sites and always made sure we got back home safe and sound.

A huge shout out to all the scientists I had the privilege to work with and learn from during the informative years of my career with the BC civil service.

The biggest thank you of all goes to Jim van Barneveld who was my guiding light for ten years – rest in peace. I have made liberal use of information from Wikipedia sites throughout the book. Footnotes are provided for statistics, species accounts, historical information, and chronologies. I have also included expanded information in the appendix.

INTRODUCTION

In a 50-year career as a field ecologist working in the Canadian wilderness, I have had my share of unique and near-death experiences. I have had the privilege to work in spectacular and mind-boggling environments throughout British Columbia, Alberta, Saskatchewan, and Nunavut. I've walked naked through rushing mountain torrents, flown in helicopters for more than 1,000 hours in wilderness settings, 'bushwhacked' in the roughest terrain and thickest bush imaginable for countless kilometres, been stalked by a black bear, battled wood ticks, and survived close encounters with hurtling logging trucks. I have had the pleasure of working with and learning from amazing resource scientists; through it all, I've gained a deep respect and understanding of western Canada's natural flora and fauna.

As I write this book, it is 2021 and our world has changed dramatically. From a biological perspective, we are in the middle of a devastating global pandemic. From a technological standpoint, the resources now available to assist with environmental field assessment are endless. They include all manner of satellite imagery, sophisticated GPS (global positioning system) devices, drones with mounted cameras, thermal imagery, computer tablets with remarkable apps (like Avenza, Theodolyte, and the OneDrive), and LiDAR (light detection and ranging) 3D imagery. With all of these wonderful tools for improving our inventory and mapping skills, the one part of ecosystem data collection that has never changed, and probably will never change, is the need for an actual scientist to be on the ground to physically record detailed information. This is my story of a life-long pursuit of learning, a never-ending quest for vegetation and soil data, and a profound awareness of the astounding natural world around us. I started out getting my hands dirty and collecting forest productivity data, along the way filling my brain with the infinite images of Mother Nature. I am still bearing witness to her wondrous spectacle today.

SECTION 1

Apprenticeship in the BC Government

The first ten years of my vegetation ecology career were indeed like an apprenticeship. I prospered under the tutelage of an endless array of experienced resources scientists, including ecologists, soil scientists, and wildlife biologists. I sucked up valuable knowledge like a sponge. Every day was a new adventure, with new experiences to be gained and appreciated.

IS THIS THE END?

At 7:15 am on a cold, dark, blustery morning in March 2016, my partner Josh and I drove to the helicopter base in Campbell River, BC. We were scheduled to depart at 7:40 on a short flight to Quadra Island, which lies across the strait, east of Campbell River. Our job for the day was to conduct a field audit of previously sampled ecosystem plots in an area of predominantly second-growth (new crop of trees after primary logging) forest for a carbon offsetting project. The solid slate grey sky was ominous above us as we loaded up the chopper and headed off. Upon landing, we grabbed our gear from the cargo hold and crouched nearby as the pilot exited with a plan to pick us up at the end of the day. The wind was cold and biting, and with a temperature just above freezing, the icy raindrops beginning to fall were a portent of the bad weather to come. It was a day for rubber caulk boots (the soles have sharp spikes to assist with walking on slippery wood surfaces and logs), protective rain gear, toques, gloves, and three layers of clothing. Josh was our navigator and showed me our planned route for the day. Our task was to visit five plot locations scattered through the dense forests, primarily Douglas-fir with regenerating western hemlock and western redcedar and resample the previously recorded field data to confirm its accuracy.

We would essentially follow existing roads and trails, with plots accessible by negotiating rough terrain covered with salal. Salal, a common shrub on the south coast, is an evergreen species with tough stems and leathery leaves. So, off we went towards our first location, approximately five kilometres away. We were dressed for the foul weather, including a Stanfield sweater (made of wool) over a work shirt, loose cotton pants, and full rubberized rain pants and top. Up and down rocky hills we clambered, across old downed (deadfall) trees, and through the tangle of sprawling salal. This was my first field trip of the year, and I felt energized out in the elements. So far, the going, although tough, was enjoyable. After about an hour, we made it to the first site – a hummocky knoll with a young (less than 60 years old) stand of Douglas-fir with scattered western redcedar.

After checking the field data (tree diameters, ages, heights, ecosystem classification and dominant understory species), we continued to another plot where we repeated the process. The walking was hard on the legs as the rain pelted down, and the temperature continued to drop. Around noon we stopped for a bite to eat, and the rest gave us a respite from the nasty weather and demanding hike. Despite our rain gear, we were both soaked and cold. Lunch was short, and we continued to traverse to the remaining plots. I soon realized in the afternoon that the bush hiking was taking its toll on me. My legs were stiffening, my feet had become blocks of cement, and my pace had slowed to a crawl. It suddenly dawned on me that at the age of 66, my days of demanding fieldwork were perhaps coming to an end. Who did I think I was kidding? The spirit was willing, but the flesh was weak and disagreeable! By 4 pm, darkness began to descend upon us. We were saturated and chilled to the bone. As I sat shivering, waiting for the chopper, I wondered if my field days were over. The helicopter and our ride home could not have come soon enough.

As I lay in a hot bath at my hotel, soaking my aching body, my mind began to wander. Where had all the years gone? Could I really be 66? What if I couldn't do fieldwork anymore? Is it time to hang up my caulks? Are my bush-whacking days over?

IN THE BEGINNING

The relentless, dry, harsh June wind forged its way eastward down the Similkameen River valley (in the Okanagan language Similkameen means 'creek which cuts its way through the flats'). It raised dust from the parched soil and scoured the town of Keremeos, turning clapboard buildings grey. It shook the fruit trees, rattled fences, and sucked the minute amounts of moisture from the dry air. Keremeos, which represents the western extent of the Sonoran Desert in Canada, is a town unchanged in decades. You could just as easily be back in the 1940s.

The temperature hovered around 35°C (centigrade) at 6:00 pm as I sat outside in the Elk Motel greenspace - a patch of lawn with two weathered slat chairs and an ancient picnic table. It had been another relentless, inferno of a day with the pale, limp air wearing us down, beating us into the ground as my co-worker Bill and I trudged across farmer's fields and through orchards. Our job was to dig soil pits (at selected locations) down to parent material (unmodified by surficial soil development processes), typically around one metre; describe the soil profile or horizons encountered and then refill the pit. We repeated the process throughout the entire workday as part of an ongoing soil survey of the South Okanagan area. The objective was to gather reliable soil information to aid in the mapping of soil units. The resultant map, organized in classes of agricultural productivity (1-7, with one being the highest), could be utilized to determine crop selection (apples, peaches, cherries, forage crops etc.) and develop strategies to improve growth rates.

I graduated from BCIT (British Columbia Institute of Technology) in the spring of 1970 as a Forest Technologist. The assignment in Keremeos was my first job as a 'green' 20-year-old with the BC Department of Agriculture. At the time, Keremeos had a population of around 1,000 (in 2016, the population was 1,373). Along both sides of the valley were steep rocky slopes with jumbled blocks of talus at their base. Talus blocks, derived from material that has sloughed off the steep, rocky terrain, can be up to 'house' size. The rock cliffs are home to Rocky Mountain bighorn sheep and mountain goats. Rattlesnakes, black widow spiders, scorpions, and cactus characterize the desert occurring to the east. It was my first field lesson – you never know what kind of creatures (or things) you will find in your study area.

Bill told me a story of working in the Fraser Valley the previous year with two partners. They had stopped for lunch in a shady grove of red alder while doing soil survey work. After settling in and starting to eat, they noticed an unpleasant odour wafting upslope. The smell was pungent, overwhelming, and persistent, so they investigated. After scouting around, they made the gruesome discovery of a long-dead suicide victim hanging from a tree. Shocked, they contacted the local constabulary. Needless to say, lunch was cut short, and work continued in a sombre fashion.

When soil surveys were being done in areas with farms and rural residences, the interface with people and animals could be interesting and sometimes challenging. There was the potential for gun-wielding landowners, snarling dogs, barbed wire, rambunctious bovines, booby traps, and curious horses. After two weeks of digging dusty pits and battling the constant wind, I was ordered back to my home base for my next assignment.

I flew to Victoria, and at the local Ford dealership, was handed the keys to a brand new, shiny 4 x 4 pickup truck and told to drive it to Fort St. John in northeastern BC. My job would be to work on forest productivity surveys as part of the CLI (Canada Land Inventory) program which had been initiated across Canada in the 1960s. The only problem was I had never driven a large pickup truck, having only recently obtained my driver's license in my dad's Volkswagen beetle and with only two months of driving experience. I merrily headed north up Highway 97 – through the arid lands of the Thompson River valley near Kamloops, on into the Cariboo through Williams Lake, then turned northeast where Highway 16 crosses the mighty Fraser River at Prince George to traverse the Rocky Mountains and Pine Pass, and down into the farmlands and vast boreal region surrounding Fort St. John. The region was home to vast areas of bogs dominated by gnarled black spruce, miles of spongy muskeg, and ever-persistent black flies. The drive was tenuous to say the least – numerous times, I found the truck rumbling on the highway shoulder, but eventually, I made it. I was working with Joe Smith (that really was his name) on CLI forestry plots. Every day we traveled dirt or gravel roads to find previously located plots. Once there, we laid out our sample plot, an area 132 x 66 square feet (two chains by one chain – a chain being an old forestry unit of measure) being equivalent to one-fifth of an acre. We measured every tree (greater than seven-and-a-half centimetres in diameter) with calipers then determined the average size diameter for every species of tree present. We then took growth measurements (diameter at breast height, age, and height) for three selected trees to represent the average size for dominant species. With the data, we could calculate the volume of dominant tree species and hence productivity. The CLI program was implemented for forestry, wildlife, agriculture, and recreation, with all lands mapped in seven productivity classes. The program was amazingly progressive for the point in time when it occurred.

The maps produced are still used to this day to assess resource capability throughout Canada.

Within the first week of work, my driving inexperience caused a problem. On the way home one day, with me behind the wheel, the truck began to slide on a gravel road, with the back end of the truck veering to the right. Unfortunately, I panicked, and my application of the brakes (totally the wrong response) worsened the slide, and ultimately the truck rolled onto its right side facing the direction we had come from on the road. Within a few minutes, enough helpful people arrived on the scene so that we could push the truck back onto its wheels. The right side of the truck was all crimped.

"Holy shit," said Joe.

"Oh damn," I exclaimed as I surveyed the damage to the side of the truck.

The explanation to my boss of what had transpired included an imaginary deer jumping in front of the vehicle. Amazingly my 'lie' saved my ass, and I was never punished for my bad judgement.

In July, the CLI forestry crews were working out of Williams Lake. Williams Lake was a rough, cowboy town where segregated bars (for whites and Indigenous folk) existed. One of the crews consisted of Paul (Nobu) Ono and John Woods. I became good friends with Paul over the summer and fall and ended up marrying his sister, Anne, in 1974. John became famous for two incidents that summer. Earlier in the year, in Fort St. John, he had made the mistake of driving a truck with a camper attached beneath an overhanging section of rooms at a motel. The result was damage to the building and a detached camper. He also caused the poor person sleeping in the overhanging room to be flung out of bed. Later that summer, when he was trying to ford a swiftly flowing river in the truck, the cab became filled with rushing water, and his project aerial photos ended up floating away, never to be seen again.

As the summer of 1970 progressed, my 4 x 4 handling skills improved. I loved my job, being out in the wild woods every day and seeing new vistas at every turn. I'm not sure how it happened, but later in the summer, I was given the chance to work with Jim van Barneveld, a University of Victoria plant ecologist, and Gary Runka, a soil scientist otherwise known as a pedologist, in the Bulkley River valley near Smithers. One memorable day we drove from Smithers northeast into the subalpine environment of Hudson Bay Mountain. The day became significant for me as a defining moment in my early career. On this summer day, my eyes were opened wide to the mind-blowing wonders of botany and plant ecology. My job was to dig soil pits, measure plot data, and help collect plant specimens. Jim would collect plants that he was unsure of or didn't know. We spent the entire day hiking and working in the alpine environment; the slopes were covered with an array of spectacular flowers. I saw brilliant scarlet Indian paintbrush, vivid yellow arrow-leaved ragwort, huge patches of deep blue Arctic lupine, and rolling beds of pink and white mountain-heather. The sun and wind beat on us all day, and when we turned for home, we were sun-burnt and wind-burned but happy. I was sad to finally leave the alpine meadows and trundle back to our meagre motel.

After dinner, I helped Jim to prepare and press the plants we had collected. They would be squeezed tight between layers of paper and cardboard, eventually becoming permanently dried for future examination and identification. Little did I know that my botanical awakening would profoundly impact the rest of my life and career. Over the next weeks, we continued sampling ecosystem plots throughout the valley. Jim would work long days, sometimes even resorting to the use of a flashlight to complete plots and long nights pressing plants until the wee hours of the morning. The botany bug was coursing through my veins now. I just wasn't aware of it yet.

That fall, I was doing forestry plots north of Fort St. John and had an experience that shook me up. I was walking along a game trail to a sample site with Steve, my assistant. Steve had grown up in Fort St. John and was an avid hunter. As we followed the path through a young black spruce forest, we came across the leg of a moose lying on the trail. We followed the trail for another 20 metres and arrived at a clearing about 15 metres across. All the trees had been flattened in every direction, and in the middle of the clearing was a massive pile of brush about two metres high. Steve, looking surprised, said very sternly, "We have to leave right now." As he did an about-face, I didn't ask why and quickly followed him back out the way we had come. As we walked, he said, "There was a cached moose under that pile of brush. Probably a grizzly had taken him down, knocking all the trees over in the process. The bear is likely really close, keeping an eye on his cache." Of course, we never did see the bear, but the images of the huge pile of debris and the moose leg stayed with me for a long time.

3 YOUNG AND IN A HAZE

After spending the winter working in the drafting shop with Sergio Bertolami (a notorious pincher of female bottoms) and Fred Waterman (charming old skier and womanizer), I was dispatched to the field in May 1971 to again work on the CLI forestry plot sampling program. Fate resulted in my field partner being Steve Lord, a UBC (University of British Columbia) Forestry student. Steve was tall, gangly, rangy, and naïve (just like me). We spent most of the summer working in the West Kootenays (including Nelson, Castlegar, Salmo, Trail, Kaslo, Beaverdell, Fruitvale, Silverton, Slocan City, and Creston) without the presence of any supervisors – toiling away in the endless sunny days and partying at night. We were in our early twenties – foot-loose, feckless, and fearless. Truth was, we thought we were invulnerable, and our actions almost resulted in catastrophic outcomes numerous times.

We had one session working with Mike Yamada, another UBC Forestry student and compatriot of Steve's. On one of our work forays, we came across a field of 'cannabis,' or what we thought was cannabis. Excited at our find, we picked a couple of garbage bags full and hurried back to our motel. Once there, we dried it in the oven and rolled humungous reefers, which we puffed away on. After smoking numerous 'joints,' we came to the sad realization that the plant we had picked was an imposter and not cannabis at all. Years later, when my plant identification skills had improved, I realized that the plant we had been smoking was a cinquefoil, a member of a large group of weedy species with palmate (hand-shaped), serrate (toothed) leaves that resembled cannabis.

Throughout the summer, Steve and I ranged all over the West Kootenays. During the day, we would locate our forestry plots and measure the trees – at night, we would smoke pot, do other drugs, or drink wine and beer. One eventful day, Steve and I drove from the Kelowna office to Nakusp and picked up an American hitchhiker on route. Steve and I had scored some 'mescaline' (it was more likely LSD or acid) and were going to 'drop' it once we arrived in Nakusp. So, after a spaghetti dinner which we shared with the hitchhiker, the three of us dropped the chemical. It soon became apparent that we were 'tripping.' I remember looking in the sink and seeing the spaghetti glowing and writhing like so many psychedelic worms. The walls of our room were 'breathing' or shrinking and expanding. After leaving the motel room and wandering along the main street, we found ourselves on a deserted beach. We all began to experience unpleasant cramps, and the hitchhiker suddenly exclaimed that the acid had been laced with strychnine (hence the radical hallucinations). Apparently, strychnine was used to transport the acid into one's bloodstream a little faster (the slang term was 'backbreakers'), although recent research has debunked this notion as a myth. I looked up, and the inky sky was filled with tracer-shooting stars. Everywhere I looked, the stars appeared to be blasting across the sky in all directions. At one point, Steve and the hitchhiker had a 'gunfight' (without guns) in the deserted main street of Nakusp. Poor Steve was having a bad trip and was seeing people with knives coming at him and mutilated corpses. Luckily for us, Jim Belsham, another field worker, was around and was able to calm Steve down over a few hours by walking and talking with him. By morning we were back to 'normal' and ready for more fieldwork.

While we were working in Nakusp, one of the other soil scientists, Horst Baender, rescued two baby squirrels that he had found when a tree fell across a logging road. He had brought them home, not quite knowing what to do with them. We immediately offered to help and took them off his hands. We named them Siddhartha and Govinda, after two characters in Hermann Hesse's book Siddhartha. Hesse's books had become popular with the counter-culture enthusiasts in the 1960s, and I was reading them. I don't remember much about looking after or feeding the wee creatures, but we raised them successfully for about a month before releasing them in a piece of native forest not far from town.

We had many adventures over the summer, including meeting Verna and Irene in Castlegar. I think it was in a local bar one night that we made their acquaintance. I eventually got lucky in the front seat of our government truck, parked in an abandoned drive-in theatre. Steve never did. Verna was a sad lady who used to regularly cry in her beer, and Irene would tease Steve mercilessly and then reject him. Irene was the granddaughter of a famous Sons of Freedom (also know as Freedomites) Doukhobor leader, Peter Verigin.

There were many Doukhobor people living in the Castlegar area of BC in the 1970s, and to this day, many of their descendants are still there. The Doukhobor movement began in the 18th century in southern Russia (details on the Doukhobor history are taken from the CBC Archives). Its followers were peace-loving people who believed that God lived within each person. In 1785, an archbishop accused them of opposing the Holy Spirit and called them Doukhobors – or 'spirit wrestlers' in Russian. They liked the name and took it for themselves, deciding that they were indeed fighting for the Spirit. Leo Tolstoy, also a pacifist, helped thousands of them immigrate to Canada in 1899. Peter Verigin was one of their leaders.

In 1924, a bomb destroyed a CPR (Canadian Pacific Railroad) train Peter Verigin was riding on. Verigin and many other passengers died, but no one was ever charged with the crime. Today the remaining descendants of the Doukhobors live peaceful lives in the

West Kootenays. All that is left of reminders of their violent history are the crumbling, ramshackle communal houses in communities like Grand Forks and Castlegar.

The Kootenays are home to a large population of black bears, and we would regularly see them during our field days. One day while Steve and I were doing a plot about 15 metres apart from each other, a bear nonchalantly walked between us as if he had not a care in the world. We were surprised but not really scared as he had not exhibited the least bit of aggressive behaviour. On another occasion, we met a pleasant young lady while travelling on the Kootenay Lake ferry from Balfour to Crawford Bay. Seeing that we were sturdy young men, she asked if we could provide her with some assistance. She related a story of a bear exploding through her cabin window at night and having to shoot it. She then asked us if we could come with her to the cabin on the Lemon Creek Road to make sure that the bear was, in fact, dead. We followed her vehicle up a winding narrow track, and after what seemed like an eternity of bouncing around in the truck, we came to her cabin in the forest. Sure enough, the window had been destroyed and lying on the cabin floor in a pool of dried blood was a massive black bear. We helped her clean up the cabin, although the bear carcass was too big for the three of us to move.

In 1971 the Viet Nam war was in full play, and the West Kootenays was a favoured location for conscientious objectors to 'hide.' The war was front and center in the news when I was growing up in Vancouver. I even remember taking part in protest marches as a teenager wearing a recycled American army jacket. The war was a tragic event in American history, resulting in the loss of about 60,000 young men in Viet Nam.

American draft evaders (typically referred to as 'draft dodgers') and military deserters sought refuge, from the United States authorities, in Canada. Between 1965 and 1975, Canada became a place of refuge for the draft dodgers and deserters. Because they were not formally classified as refugees but were admitted as immigrants, there is no formal count of how many draft evaders and deserters came to Canada during the Vietnam War. One estimate put their number between 30,000 and 40,000.

These young men, who were typically well-educated, had a significant impact on Canada's academic and cultural institutions and Canadian society in general. While some draft dodgers eventually returned to the United States, after a pardon was declared in 1977 during Jimmy Carter's presidency, approximately half of them stayed in Canada.

One day, when Steve and I were driving down a rough, narrow road into the hills west of the Slocan Valley, we came across a 'commune' in the middle of nowhere with 'hippies' (and likely, draft dodgers as well) dancing naked in the fields. The commune consisted of a few ramshackle buildings and some teepees. A couple of flaxen-haired, young women came running over to our truck, giggling and smiling at us. They asked us if we would like to smoke a joint with them (they were obviously high), but we had to decline since we were working. However, we did visit with them for a while and have a cup of herb tea. They told us they were students from Vancouver and had come to the Slocan to get away from the urban rush and get back to nature. After enjoying their sweet company, we sadly departed and continued on our way down the winding road.

The Slocan Valley was a favorite place for the draft dodgers to settle due to its relatively isolated environment and lack of much vehicular traffic. Highway 6 runs north through the Slocan Valley, connecting Highway 1 (the Trans Canada Highway to the north) and Highway 3 (the Crowsnest Highway to the south). From its southern starting point at Pass Creek to its terminus at Nakusp, Highway 6 extends for close to 200 kilometres. The valley was also home to many of the internment camps for the 22,000 Japanese Canadians who were forcibly uprooted, dispossessed, and incarcerated during the Second World War[1] (see Footnotes and

Appendix for an expanded discussion of the Japanese internment camps).

In the fall, Steve returned to Vancouver for school, and I went back to the office in Kelowna. Looking back on the summer of 1971, it was amazing to me that neither Steve nor I suffered any serious injuries during our reckless field season. I don't recall receiving any safety training; I'm pretty sure there was no first aid kit with us, we never communicated to anyone where we were going. We merrily travelled down active logging roads without a radio, and we were stoned almost every night after work. None of those actions or activities would ever be condoned in today's super safety-conscious work environment. I think we both had angels on our shoulders that summer.

As it happened, Jim was hired as a permanent plant ecologist in the fall of 1971 and needed an assistant. I was asked if I would be interested and jumped at the chance. My initiation consisted of reading a plant taxonomy manual then working my way through stacks of pressed plants to try and key them out (identify them). I learned how to differentiate families from each other based on flower structure and form and the common genera within families. It was a challenging process that continued for months over the winter and spring, but by the 1972 field season, I had a rudimentary understanding of plant identification and plant ecology. Armed with this basic understanding and my own innate curiosity, I was all set to start doing my own ecosystem plots.

4 BUDDING BOTANIST

In 1972, under the direction of Jim, I was tasked with collecting ecosystem data within the Nelson map sheet (82 F at 1:250 000 or four-mile sheet). Ecosystem plots were selected to characterize the variation in parent material (how surficial material was deposited), soil development (soils develop horizons in response to regional and local changes in climate and material), and vegetation within the entire map sheet. Essentially, I spent the summer working in the same general area that Steve and I had patrolled the year before. I still had a strong forestry bias, so to start with, my plots were often chosen on the basis of forest productivity and not ecology. Still, as time passed, I became more focused on ecological traits such as how the stand originated (cleared or burnt), dominant growth form (trees, shrubs or herbs and grasses), aspect (the direction the site faced), elevation, and parent material.

My first field assistant that summer was Gary, who turned out to be a social miscreant lacking the minimum of social graces and virtually no driving skills. Gary had grown up with an abusive father and alcoholic mother and had developed a challenging personality. Invariably every morning, Gary would leave a large fecal deposit in the toilet to greet me; he had no concept of food preparation, continually sucked on his beard, and was so shy he could not look me in the face while talking.

Gary's driving made me very nervous – to the point where I would be looking back and forth between the vegetation passing by and the road ahead of us. On one occasion, while driving down a rough gravel road outside of Kaslo, he failed to see branches with pink flagging tape sticking out of a large hole in the road. Before I could stop him, he drove the truck's front end into the massive sinkhole, breaking our front axle. I was furious and started chewing him out for being so stupid. There were no cell phones in those days, so I ordered Gary to run back to Kalso (at least five kilometres) to arrange for a tow truck. He did this once he stopped crying. After a couple of weeks of tolerating Gary's behaviour, I decided that maybe sharing a joint with him might help him relax. So, one night in our motel in Nelson, I offered him a few puffs. We were drinking beers, and I playfully flicked a beer cap at him. A strange look came over his face, and he got up from the couch, walked over to me and poured his beer on my head. Somehow, I calmed myself down enough not to punch him and walked out of the motel to cool off. On my walk, I found a pay phone. I phoned Jim and told him that Gary was not going to work out and get me a new assistant.

One of my assistants that summer was a strapping rugby player by the name of Steve Thompson. In later life, Steve served as an MLA (Member of the Legislative Assembly) for the riding of Kelowna-Mission. One night while camped near Salmo, we were pestered by a marauding black bear. We could see and hear the bear roaming around as we slept in the back of our truck. At one point, the people in the campsite next to us abandoned their tent and sat huddled in their car as the bear ransacked their belongings. The bear then moved to our site. Steve was not going to tolerate this intrusion and leapt from the back of the truck in his skivvies to hurl a large boulder at the bear, just missing his head. It was probably lucky that the boulder didn't make contact with Mr. Bear. After a fitful night, we headed off to work in the morning, fully aware of the bear huffing and puffing nearby in the trees. Upon our return at the end of the day, we found our belongings, including soil samples stored in coolers, strewn all about. I came to learn over the years that 'campsite' bears were amongst the most dangerous, having become habituated and addicted to human garbage and food.

Over the course of the summer, my plant identification skills improved. I continued to collect, identify, and press plant specimens every day. Because scientific (= Latin) names are standardized regardless of jurisdiction, and common names are not (between provinces, for example), I was required to learn the scientific names of all plants. Each plant species had a genus and specific or species name. For example, devil's club was *Oplopanax horridus*; bunchberry was *Cornus canadensis*; bastard toadflax was *Geocaulon lividum*. It was challenging to learn the plants this way. I developed a system whereby I would focus on learning ten new plant species every day. While 'describing' the ecosystem being sampled, I was required to make a list of all the species I could recognize, including trees, shrubs, herbs, grasses, mosses, lichens, and liverworts, as well as estimate the areal coverage that each species occupied. The variety and diversity of vegetation within our study area was astounding – arid grasslands on dry, warm slopes; ancient climax forests of western hemlock and western redcedar; seral forests of Douglas-fir, western larch, paper birch, lodgepole pine and trembling aspen; subalpine stands of Engelmann spruce and alpine fir; burnt scrublands of ponderosa pine and saskatoon; and rocky alpine tundra. As time passed, I continued to expand my cerebral database of plants I could recognize. However, I was still a botanical neophyte and really had only a minimal understanding of how it all fit together.

5 THE FOREST LOOKOUT

The 1973 field season found me working in the 82L map sheet (Shuswap); the area covered transitional (from dry to wet) environments, western hemlock and wester redcedar forests, subalpine and alpine domains, and even grasslands bordering on desert. Towns within the study area included Revelstoke, Chase, Salmon Arm, Vernon and Sicamous. As usual, our crew utilized the back of our truck as sleeping quarters. Our crew consisted of Andrew Harcombe, Ted Lea, Mohammad Rafiq, Eric Stathers, Rick Ferster, and I.

To reiterate, it was amazing how ignorant and naïve we were back in those days. We merrily hurtled down active logging roads without radios, worked in bear country without bear-awareness training, and travelled to isolated areas without advising anyone of our plans or anticipated whereabouts. Today, a resource worker cannot set foot out of their truck, lift a shovel, or start any work process without successfully completing a myriad of safety courses. Those safety requirements are predicated by risk and liability issues for all employers mandated by government legislation and insurance companies.

In those days, a big part of fire detection involved a network of forest fire lookouts built on mountain tops throughout BC with panoramic vistas of the surrounding landscapes. They were typically manned by a hardy individual who spent the entire fire season being vigilant and lonely. In 1973 there were 320 forest lookout towers in BC (many were established prior to 1940); today only 20 remain.

Since our primary access to sample sites was by backwoods roads, often in very rough condition, we occasionally travelled up forest lookout roads. On such a day, Ted and I left Enderby. We headed east along the Shuswap River, with our intended destination being the Mara Mountain Forest lookout, a drive of approximately 25 kilometres. It was a clear sunny morning in July, and we followed the Shuswap River until it turned north toward Sugar Lake, where we found the signpost to Mara Mountain (1,155 metres) and headed up a narrow gravel road.

Over the next hour, the road deteriorated rapidly so that at an elevation of 1,000 metres, we were in 4 x 4 mode and crawling through washouts and along a boulder-strewn trail. We had left the Douglas-fir and lodgepole pine forests behind and now passed into subalpine fir, and Engelmann spruce stands with widely spaced trees and a dense layer of white-flowered rhododendron beneath. On our way up the rubble-strewn track, we came upon a small, herbaceous meadow hidden in a low swale. A stunning array of gorgeous flowers greeted us in full bloom – brilliant heart-leaved arnica, white Canada anemone, mountain mugwort, pink mountain-heather, and purple sticky geranium. Ahh, and what was this? A tall (greater than one metre) robust, blue plant with irregular showy flowers arranged in a magnificent spire. It was unknown to Ted and me – what could it be? Ted and I pulled out a field guide and thumbed through it. Eureka – there it was. Tall larkspur, a member of the fascinating Buttercup family. It had large flowers with parts of it white and portions dark blue. Like many buttercup species, all parts of the plant are toxic to humans and livestock from the alkaloid methylcaconitine. If ingested, death could occur with complete motor paralysis and acute respiratory failure.

We continued crawling up the ever-worsening road (now more like a dry creek bed), catching occasional glimpses of the sprawling panorama below us. Eventually, we emerged into a parkland-subalpine meadow complex with unparalleled vistas in all directions. Black flies and horse flies were swarming around the truck, and as we disembarked, the flies harassed us mercilessly. Just ahead of us, we could see the forest lookout standing atop the mountain.

This next section is contrived only to show how the lookout person would live on a day-to-day basis. I obviously can't remember his name, but the account gives some idea of that way of life in the 'good old' days.

As we climbed the stairs of the sturdy 30 metre high, white-painted structure, the lookout man came out to greet us. "Hello below," he yelled. "How are you guys this morning? Can I interest you in a cup of tea?"

We replied with a hearty yes and followed him into the lookout. Introductions followed - his name was Mackenzie (Mack) Wilson. "Welcome to my humble abode," he said, laughing. "It's not too big but has everything I need."

The living space was indeed small, only about 25 square metres, but exceptionally tidy. On one side of the room was a small bed with a nightstand, a bookshelf and a chest of drawers; on the opposite side was a continuous worktable with topographic maps, aerial photographs, rulers, and pencils laid out. Along another wall were the propane stove, sink, and small fridge. In the middle of the room were the fire spotting stand and a large map. Mack showed us how to use the 'fire-finder' to locate smoke and record azimuth degrees (above and below the horizon) and distance. Large clear windows surrounded the entire room, while the outside had a continuous walkway with a hand railing. Water was stored in a large holding tank at the base of the tower, and power was supplied by a small generator.

Mack chuckled, "So what brings you up to the top of Mara Mountain on this fine summer morning?" I described to him what we were doing, and our survey methods and could tell he was a keen listener. I asked him what his regular daily routine entailed.

Mack said, "You know, my days go by pretty slowly. I rise at daybreak, anywhere from 5 to 6 am depending on what month it is. I'm here from May to September, with occasional breaks to spell the monotony. Once I have dressed, washed, and eaten, I do a full 360 degrees scan with binoculars, looking for smoke. Any smoke that I observe has to be reported to the Enderby Forest Station via radio. I hike around a little bit, but not too far from the lookout, and I read a lot and do some birdwatching. I also write a bit. Just short stories and poems. I'm also responsible for recording the climate data in the morning and evening."

The kettle started to whistle, and he quickly prepared our tea which we eagerly sipped. After an hour of listening to Mack recount stories of occasional bears passing by, inebriated hikers, and violent thunderstorms, we prepared to leave. Just before doing so, he confided in us the location of his 'secret' fishing spot. With a wink, he said, "Easy to catch, endless trout. Just down the road a bit and hang a hard, right turn. You'll have a steep hike down and up, but well worth it."

We climbed back down the stairs and hiked back to our trusty truck. Bumping downhill for about a km, we came to the pullout that Mack had described. We grabbed our rods and clambered down a steep, narrow trail to a small jewel of a lake with a clear azure-coloured countenance. Since it was near noon, we figured the lake would be a good lunch spot. The 'fishing hole' was so clear and shallow we could see trout cruising everywhere. On Ted's first cast, he hooked a fish and quickly landed a beautiful, speckled trout about 30 centimetres long. After a few casts, I also managed to land a lovely fish. We caught a couple more and then stored them in our lunch cooler for later consumption. We continued our work through the day, always winding downward into the valley below us.

The summer full of bright, shiny days and little rain passed quickly. My plant identification skills were growing in leaps and bounds.

AN EPIPHANY

In 1975 our biophysical inventory program took us to the East Kootenays region on the eastern edge of BC. The Columbia River valley, also known as the Rocky Mountain Trench, resulted from the passage of the Columbia River flowing north from Columbia Lake following the deglaciation of the landscape. To the north, the valley becomes an expanse of marsh wetlands up to two km wide before turning south (north of Revelstoke) and travelling a total length of 2,000 kilometres to the Pacific Ocean near Astoria, Oregon. The drainage basin, 668,000 square km, is larger than France, and the river has been used extensively for travel and trade since the early 18th century. Power generation began in 1942 with the massive Grand Coulee dam located in Washington state (the sixth-largest hydroelectric dam in the world). The river is bounded by the Columbia Mountains to the west and the Rocky Mountains to the east.

Our surveys took us high into those mountains, and on one of our trips to the alpine in July, Rick Ferster and I drove for hours before arriving in the subalpine forests of Jumbo Mountain. We shouldered our backpacks with lunch, rain gear, field equipment, and water and began hiking upwards. Our destination was the alpine meadows at around 3,100 metres, reputed to have spectacular wildflower displays. Around noon we arrived at impressive meadows full of lupines, fleabane, false hellebore, sedges, and monkey flower with gurgling snow-melt freshets singing a happy early summer song. I said to Rick, "Let's try and get to that scree slope above us; there could be some good fossil hunting." Huffing and puffing as we scrabbled upward, I suddenly stopped dead in my tracks. Quietly I whispered, "Bears at 12 o'clock." I heard Rick audibly gasp as a huge sow grizzly with two cubs emerged from a thick patch of willow about 100 metres away. Luckily, the trio didn't spot us as we stayed perfectly still, and they continued away from us downslope, foraging as they went. Finally, feeling safe after they disappeared from our view, we continued our trek to the scree field.

As I glanced down, I couldn't believe what I saw. "Wow, Rick, look at this!!" I exclaimed as I reached down to grab a peculiar-looking and clearly ancient rock. The face was composed of numerous tubular structures, which I later discovered were fossilized homes of sea worms (we also found fossilized coral) dating back 500 million years to the Cambrian period. Here before us was clear first-hand evidence that the Rocky Mountains had formed under oceans millennia ago. As a neophyte scientist, I was blown away by the real-life paleontology history lesson, one that stayed with me my entire life.

Rick and I were part of a ten-person field crew that summer. Our initial work began with a correlation and training session with Jim orchestrating the proceedings. Most of the crew were university summer students with little or no field experience. It was on one of our training days that I experienced an epiphany. Jim had been explaining the terrain processes (how modern-day landforms [vis a vis the landscape] came to be), and even though I knew that the glaciers had melted beginning about 12,000 years ago, I didn't really fully grasp the big picture. I couldn't really explain why, but as we were travelling in the valley bottom near Skookumchuck, it all suddenly crystallized for me. I saw the light, and I truly could 'see' and understand the landscape from a surficial material perspective. As the glacial ice that filled the Columbia River valley (the ice would have been up to 1.6 km thick) melted, numerous depositional processes occurred. As the mountain tops cleared of ice resulting in massive flows of water (called glaciofluvial), the torrents carried a mixture of gravel, boulders, sand, clay, silt, and blocks of ice downslope. Because the valley floor would have still had considerable ice present, the glacial meltwater deposited its loads along the sides of the ice against the valley walls forming terraces. The terraces, once you knew what to look for, were easy to see, and now I could. This is a simplistic explanation of only one type of deposition. Surficial materials come from a variety of sources and are deposited over the landscape in different ways (gravity, ice, water etc.). The types of material, texture, and size have a direct bearing on things like soil drainage, slope stability, and water retention. This knowledge of how landforms were created and how the materials behaved was a cornerstone of determining relationships for ecosystems. That was my epiphany.

7

THERE'S BLACK GOLD IN THOSE HILLS

In 1976 the RAB (the Resource Analysis Branch for whom we worked) took on the challenge of conducting biophysical surveys for the Northeast Coal Project. Our ecology team included Del Meidinger, Jim Fyles, Andrew Harcombe, Phil Comeau, Anne Comeau, Cliff Dyer, Terry Wood, Wayne Blashill, Maureen Ketcheson, and I. Our task was to collect vegetation and soil data in a vast study area within which three coal mines were planned for Bullmoose Mountain, Quintette Mountain, and Sherriff Mountain. The mines would all be developed for coal extraction in the northern Rocky Mountains. The nearest town was Chetwynd, and we worked out of there as well as Dawson Creek and Beaverlodge, Alberta. As well as the spectacular Rockies, the area included the future townsite of Tumbler Ridge, the Kinuseo River, and the Pine River with extensive tracts of boreal forest, sweeping alpine slopes, and rich wildlife habitat for a multitude of species.

The project was envisioned and promoted by the Social Credit government of the day, which built a $400 million, 300 kilometres long rail link to Prince Rupert to allow for coal to be transported to Japanese buyers. The ultimate total investment, including federal funding, was around $1.6 billion, a massive amount of money for 1976. The mines were closed in 2014 due to a crushing downturn in the coal industry but were being operated in 2019 by Conuma Coal.

Nearly all the crew were recent graduates of the University of

Victoria with limited field experience. A case in point was the 'night of the wild horses' experienced by Wayne and Maureen. Wayne, Maureen, Cliff, and I had driven into the Bullmoose Valley and set up camp with two tents. Our camp was in an open area near a road ROW (right-of-way), and after dinner and fireside chats, we turned in for the night. Unbeknownst to any of us, a herd of wild horses used the right-of-way as a travel route, and their periodic galloping went on all through the night. Cliff and I heard nothing, but poor Wayne and Maureen got little sleep, terrified by what they thought were rampaging moose. To add insult to injury, a poor toad was trapped under their tent all night long. Wayne and Maureen went on to have long careers as resource scientists in BC. 1976 would be their year of 'learning the ropes' in ecosystem sampling. The Northeast Coal Project was the first time we used a helicopter extensively to access wilderness areas. A typical day would see our pilot, Bill Carr-Hilton, moving crews to various transect locations with his four-passenger Bell JetRanger 206. The JetRanger, which was introduced in 1967, is still being produced and used today. Our transects, typically running from mountain top to valley floor, would consist of collecting soil and vegetation data at selected locations throughout the day. Bill would pick each crew up at a pre-selected pickup site and transport us back to camp. Tragically, Bill was killed in a car accident after the project completion; Mount Carr-Hilton in northern BC is named in his memory.

The study area was home to a wealth of big-game animals including cougar, black bear, grizzly bear, lynx, bobcat, wolf, wolverine, moose, Rocky Mountain bighorn sheep, mountain goat, woodland caribou, Rocky Mountain elk, and deer (both white-tailed and mule). Maureen was terrified of bears and dreaded the long transects through the rich, verdant areas of bear habitat. We all wore small bells on our backpacks or cruiser vests and regularly yelled as we travelled through the bush (we didn't have bear spray back in those days). Today we carry defensive bear spray used for warding off potentially hostile bears, but in those days, we had no such thing. I had many bear encounters over the years, and the bell saved my ass a few times.

The flora of the northern mountains was diverse and rich, and as such, Maureen and Wayne would invariably bring back bags of plants to be identified or placed in the plant press. During one such evening, poor Maureen began wailing loudly, "I can't take any more of these f---ing plants!!!" pulling out a tuft of her hair and throwing a pile of specimens on the floor. We were all challenged to learn as much of the flora as we could in a relatively short time, and over the next few months, we became reasonably proficient. None of our field workers were mauled by bears during the field surveys, and ultimately, our data and maps of biophysical units were utilized to reconcile the impending mine developments with the habitat needs of the big-game animals.

ADOLF AND THE WOLVERINE

In 1977 our group was tasked with completing biophysical surveys in an extension to the original Northeast Coal Biophysical area. As part of that work, we engaged Adolf Ceska, an eccentric and endearing botanist. Adolf was terrified of bears, to the point where he generally wouldn't do work in areas where he knew his risk of running into a bear was high, although this included pretty much all of the wilderness areas in BC. Terry Woods was assigned to be Adolf's field assistant. Adolf's main interest was botany; as such, he collected overwhelming amounts of plant specimens to be placed in his plant presses for later identification. I remember visiting Terry and Adolf where they were staying in McBride (about 200 kilometres southeast of Prince George, down Highway 16). In order to dry his plant specimens efficiently, Adolf had rigged an interesting contraption to his plant press. He had a hairdryer attached to the press, which pumped a continual supply of hot air to the specimens. Normally, specimens take at least a week to dry. With Adolf's ingenuity, he had reduced the drying time to a day or two.

During the summer of that field season, Adolf and I participated in a soil correlation tour in the study area. On these tours, a large group of soil scientists and ecologists would get together for a field trip which consisted of driving down access roads to examine and discuss soil conditions and ecosystems. At one such stop on our tour, high in the mountains, we were all gathered around a soil pit discussing soil classification, soil texture, and the associated plant assemblage. Adolf had wandered off over the alpine meadows nearby, and we forgot about him. Suddenly we heard frantic yelling and bellowing off in the distance. Adolf was running as fast as he could, yelling, "It's a bear! It's a bear!" He was tearing back towards us with a wolverine in hot pursuit. Around his neck were his camera and binoculars, which bounced up towards his face with every step. Finally, after covering about 100 metres, the wolverine gave up, but Adolf took no chances and kept running towards us. The image of Adolf running across the meadow with the wolverine after him stayed with me for years. I don't think Adolf was ever accosted by a bear, and I am certain he never saw another wolverine[2] (see Footnotes and Appendix for a detailed species account of the wolverine) in his life. In my 51 years of bush work, I have only seen a wolverine on three occasions.

GALIANO NIRVANA, LEVEL MOUNTAIN, AND THE HORSERANCH

In 1978 the branch undertook a biophysical inventory of Galiano Island, one of the larger gulf islands located in the Strait of Georgia between Vancouver and Vancouver Island. The work, which was being done for the Islands Trust organization, included data collection for ecosystems, soils, wildlife, recreation, aquatics, heritage values, and visual perspectives. The Gulf Islands have a relatively mild, dry climate with an average annual precipitation of less than 500 millimetres and unique vegetation that can tolerate the dry, harsh conditions. Many of the species, such as Pacific madrone, hairy manzanita, and yerba buena, are native to the dry American Pacific coast (California and northern Mexico), and the islands represent their northern extent. Galiano Island was (and still is) home to an eclectic mix of hippies (or 'shrubs' as I call them), retired folk, and part-time residents. Its eroded sandstone cliffs along the western shore, called tafoni, were caused by salt-weathering and physical processes. They are visually striking and unique to this part of BC.

The biophysical team included a myriad of disciplines and scientists (myself – vegetation and soils, recreation – John Block, surficial geology – Alan Pattison, aquatics – Tom Chamberlain, visual aesthetics – Roger Horner, wildlife – Andy Stewart, archaeology – Stephen Cassidy). Getting around the island was easy on a good road network, and we also used Zodiac inflatable boats to access the shoreline and isolated beaches.

It was the best of times for our sampling program – cruising on the waters of Trincomali Channel and the Strait of Georgia in constant sunshine, hopping out on deserted beaches, and gathering soil and vegetation data of the dry island ecosystems.

One night, Allan, John, Roger, and I were indulging in some good weed. It was late in the evening when we were discussing the effects of microclimate on vegetation cover. I maintained that you could experience the effects of microclimate, specifically cold air ponding in landscape depressions. I told the others of a small wetland I had found in a hollow near the summit of Mt. Galiano, which overlooks Active Pass. So off we went, four stoned fools in our truck driving down empty roads, eventually turning on to the road to Mt. Galiano. Ultimately, in pitch-black darkness with me driving, we arrived near our destination. Having the foresight to bring two flashlights, the four of us, laughing and still high, stumbled around in the thick shrubbery. The coarse, leathery leaves of the waist-high salal impeded our progress, but finally, we fell into the small sedge wetland. The thick vegetation was indeed damp, and the air was markedly colder than that of the surrounding landscape. Allan, Roger, and John agreed with my theory, and we didn't linger long under the inky, star-studded night sky. We soon left the little fen and drove back to our cabins. The rest of the crew were none the wiser to our midnight sojourn.

One day, the entire crew ended up on a pebbly beach devoid of land access (we had arrived there by Zodiac for a convenient lunch stop). Stephen Cassidy, the archaeologist, explained to us that it was not uncommon to find artifacts scattered on beaches such as this. I was keen on finding something, so began to slowly move about, hunched over and looking at the array of small, multi-coloured pebbles. After about 20 minutes of hunting, I noticed a small object different from the pebble matrix. I picked it up and initially thought it was a fragment of purple glass, but on closer inspection, I realized that it wasn't glass, but I wasn't sure what it was. It was about three cm long, two cm wide, and two to three mm thick; oval in shape with one straight end and a sharp outer edge; and purple coloured with black flecks. I was excited about this 'weird' object and showed it to Stephen. His eyes widened as he held it up to the light and turned it over in his hand. "Wow," he exclaimed. "This is very cool and quite special," he said. He gave me an explanation based on his professional perspective: "What you have here is a hide-scraper, probably in the order of 10,000 years old. It's composed of chalcedony, a type of purple obsidian. It would have been transported here from somewhere in the Interior of BC and likely was offered in trade to a coastal First Nations member." He showed me how the cutting edge exhibited evidence of being 'knapped,' the shaping of a piece of stone by striking it, breaking off small portions to create a cutting edge.

He demanded that I turn the piece over to him, but I refused since I knew that the scraper being isolated on the beach and removed from a time sequence (chronological profile) had no real relevance to him. I was stunned and thrilled by my good fortune and blown away by this ancient tool. I still have it, and when I hold it between my thumb and forefinger, it fits comfortably and is warm to the touch. You can almost imagine it being used to remove the skin from an animal hide. The moral of the story was: I never knew what I might find on a deserted beach, and there were lots of them on the west coast of BC.

Galiano Island was home to poison oak (the only other place it is known from in BC are the seashore cliffs near Victoria), and I played a mean trick on the crew. Unlike poison ivy, which is somewhat common in the dry interior of BC, poison oak is predominantly a species of Ontario. On Galiano, I found it growing on large, sandstone blocks at the base of cliffs falling into the waters of Trincomali Channel. I knew what it was when I found it but didn't tell any of the crew members. At our daily evening debriefing meeting, I casually passed it around and asked if anyone could identify it. Poison oak affects people with oils that are exuded from the plant and cause mild to severe reactions (contact dermatitis). Poison oak, like poison ivy, contains urushiol. This oily substance is what causes a poison oak rash, and it can be almost impossible to avoid. Upon contact with one's body, urushiol immediately forms a chemical bond to the skin and can cause a severe allergic reaction. It can stay on clothes, pets, or other materials for months, and its potency endures. I had tested myself and didn't seem to have a reaction, but it was mean of me to pass it around the group. When I told them what it was, they were quite upset with me. It showed how thoughtless I could be back in those days.

Later in the summer, I was appointed as the plant ecologist for a unique wildlife habitat project in northwestern BC. Rick Page, a biologist with his Ph.D. employed by the Ministry of Forests, was studying the interrelationships of caribou and wolf populations. His study area included Level Mountain, northeast of Telegraph Creek and the Horseranch Range near the Yukon – BC border. Level Mountain (2,166 metres near the peak) is an ancient shield volcano characterized by subdued slopes and continuous cover of bog birch interspersed with subalpine meadows. The complex of vegetation is perfect habitat for woodland caribou from spring to fall.

In August, we prepared for the first phase of work which involved travelling to Dease Lake by truck, then catching a helicopter to take us and our camping gear to a site below the summit of Level Mountain. Our crew consisted of myself, Rick, and Craig Bissett, a summer field assistant. We planned to be working out of our camp location for five days with no radio contact, living in tents, and conducting surveys on foot in different directions. We watched the chopper disappear as we took stock of our selected camp location in a small grassy opening beside a small, babbling mountain stream.

It was romantic to be camping out in the wilderness and making our own meals by an open fire, but the truth was, we were in grizzly country and had to be wary at all times. Not only that, we were in the stomping grounds of Sheslay Free Mike. He was an American renegade trapper whose real name was Michael Oros, who lived alone in the bush from about 1978 to 1984. He was described as deranged and dangerous, and often 'borrowed' things from back-country cabins. In 1984, Oros ambushed a tactical RCMP team that had travelled from Whitehorse, Yukon to Teslin Lake to arrest him. He killed Corporal Michael Buday before being gunned down himself. Following that, a coroner's jury determined that Oros had killed a fellow trapper, Gunter Lischy, sometime around 1981. His bones had been found near Oros' cabin site by pure luck after an original search, when Lischy first disappeared, didn't find any evidence of a crime.

On our first hike out through the continuous cover of bog birch (a low shrub less than two metres tall with small oval leaves and crystalline glands on the stems), we came across fresh wolf scat, which Rick carefully examined. He told us to take great care with any wolf scat since disturbing the scat could result in the inhalation of eggs from the parasitic tapeworm *Echinococcus granulosus*. He bagged the specimen, which appeared to contain bone fragments, for future examination in the office. We were lucky with the weather, absence of bear encounters, and an abundance of stunning flower specimens. One particularly striking species was felwort, a tall (up to 50 centimetres), showy plant with large blue flowers. At a few locations, I also documented the occurrence of a rare species, yellow marsh saxifrage (characterized by having two joined ovaries [seed chambers]). After five days of alpine forays, sampling numerous ecosystem plots, and fording ice-cold streams, we dismantled our camp. The helicopter was returning for us on a pre-arranged date since we had no radio. Without communication, it took a few passes for the pilot to locate us. Grubby, unshaven, and stinking of campfire smoke, we were happy to be returning to civilization.

Once we were back in Dease Lake, I realized that I had left a good field jacket hanging on a tree in camp. Two days later, on a reconnaissance flight over Level Mountain to view and estimate caribou numbers, we had an opportunity to set down at our camp location and retrieve my jacket. Upon searching for it, I discovered that it was not where I had left it and nowhere to be found. There was only one explanation: Sheslay Free Mike had been 'watching' us from afar. He would certainly have noticed our campfire smoke, and simply took the jacket once we left. Perhaps we were fortunate to have not been slaughtered while we slept!

In early September of the same year, Jim van Barneveld, Rick, Mike Fenger, and I travelled by plane to Watson Lake in the southern Yukon to prepare for collecting additional field data on the Horseranch Range. The Horseranch Range is a 40 kilometres long, north-south oriented range of connected mountains approximately 30 kilometres south of Watson Lake. We caught a helicopter and once again camped in the alpine. Once the helicopter deposited us, as well as our gear and provisions, we set about erecting our camp. It was late in the day and the sun was sinking fast as we put up our tents in the narrow valley. The four of us were sitting around our cheery campfire talking about grizzly bears as we were once again in prime grizzly habitat when suddenly Jim casually said, "There's one now." Mike and I jumped up to look where Jim pointed, and sure enough, we could see the clear, hulking silhouette of a massive grizzly moving on a slope no more than 300 metres away.

"Holy shit!!!" exclaimed Mike as I felt a shiver of trepidation run down my spine. So, there we were in the lair of the bear, and Mike, who had bought along a rifle, began to load shells into the magazine. In response to the bear sighting at close proximity, we decided to keep the fire burning and take turns tending it. After a few hours and a few drams of Scotch whiskey, that plan was abandoned, although Mike spent a nervous night in a fitful sleep with the nose of his gun protruding from his tent. Of course, no grizzly appeared that night. In the morning, on examining our campsite, we saw obvious signs of grizzly presence, including tracks and scat.

That day, and for four more, we headed out into the spectacular alpine landscape around us. The alpine bearberry, a low ground-covering shrub, was showing brilliant colours of orange to crimson, and on the highest wind-swept, dry ridges, we found pink cushion plants with subtle blooms (another rare plant species for BC). One day as we worked our way up a warm ridge, Rick noticed some movement on a distant willow-covered slope. All he said was, "Bear." and our ears perked up. Scanning the slope with binoculars, we took turns watching a sow extricate a marmot to share with her two cubs. Great piles of soil flew behind her as she dug furiously. Sure enough, within a few minutes, she had secured the helpless rodent. We felt lucky to be observing her from a safe distance.

The rest of the trip passed uneventfully, and we ended up exhilarated, wind-burnt, and satisfied with the results of our survey. However, on the last day, we had a disconcerting experience. We were waiting late in the afternoon for the chopper at our second camp location when we heard a gunshot. The shooter couldn't be far away, and we wondered who could possibly be here in the mountains with us. Soon enough, we had our answer. Within half an hour, we nervously watched as two individuals approached our location. Upon arrival, we could see they were two First Nations men dressed roughly, carrying backpacks and a rifle. They told us of being 'fired' by a guide-outfitter who had left them to find their way back home on foot. The gunshot we heard was the sound of them killing a caribou (they removed the heart and liver to appease their rampant hunger), as evidenced by the blood on their boots. They had already trekked about 50 kilometres or so through the bush, heading for the nearest road, which lay some distance to the east. I offered to let them use one of our tents, as long as they promised to leave it when done for us to return and pick it up. We were sceptical about ever seeing the tent again, but sure enough, when we returned two days later, there was the tent rolled up right where we had left it.

SECTION 2

Pedology Consultants Ltd.

I only spent two years with Pedology Consultants, but the experience of working for a consulting company was quite different from my years in the government. I quickly learned about meeting client needs in a timely and efficient manner. As well, I was exposed to a wide range of different projects, which taught me how to adapt to the task at hand and learn new skills.

10 CANSTAR CHAOS

In 1981 I left the government job in Kelowna and took a position with Pedology Consultants Ltd. in Victoria, BC. The company was owned by Mark Walmsley, a former soil scientist with the Resource Analysis Branch, as well as Len Knapik, Al Twardy, and Len Leskiw, three soil scientists in Edmonton, Alberta.

My first big assignment was to be the vegetation ecologist on a biophysical inventory program for the Canstar tar sands (also called oil sands) project north of Fort McMurray in northern Alberta. The study area, which was only accessible by air or by taking a barge across the Athabasca River, was part of the vast Boreal Region characterized by extensive black spruce bogs, tamarack fens, dry jack pine ridges, and valuable deposits of bitumen. Bitumen is composed of hydrocarbons that were formed hundreds of millions of years ago from the decomposition of algae and other marine sea creatures. At certain depths and temperatures, this organic matter decomposed to form oil. The oil is believed to have been formed from inland oceans which covered much of western Canada. Formation of the Rocky Mountains pushed this oil to the east towards the Alberta and Saskatchewan border, transporting with it, water suffused with oxygen. Bacteria contained in the water fed off the lighter hydrocarbon molecules in a process called biodegradation. As a result, a heavier complex hydrocarbon, containing sulphur and trace concentrations of metals, was left behind to form Alberta's oil sands deposits.

Our home base was a construction camp consisting of mobile dorm trailers and a large dining room. The project involved collecting ecosystem data (vegetation, soil, and wildlife) to meet the regulatory requirements for approval to begin mining. Our crew consisted of myself, Craig Heath, Len Knapik, Larry Nikiforuk, Dave Westworth, Briar Edmonds, and Kathryn Bessie. Seismic lines riddled the work area, and we accessed sample sites on three-wheeled 'trikes' and with a helicopter. The weather that summer was unusually hot and dry. Because we were often wading through wetlands with deep water, sometimes up to our waist, I gave up on wearing work boots and resorted to sneakers. It was so dry that yellow jackets would often alight on my shoes to extract moisture. Late in the summer, there were forest fires burning all around the study area, and the air was full of smoke.

Temperatures were often at or above 30°C, and without air conditioning, the camp trailers were like saunas by the end of each day. There were limited safety standards, and the concept of a 'dry camp' didn't exist in those days. After time off, crew members would return to camp with suitcases loaded with bottles of booze, and we would regularly have pot-smoking and drinking parties down by the Athabasca River. Camp was a never-ending gong show. One morning a full-blown food fight broke out. In the middle of the night, a crew member came across a burly truck driver having sex with the cook's daughter on the floor of one of the washrooms.

We often played football or Frisbee in the camp parking lot after dinner. I was in my room one night, having puffed on a joint, watching a game of football when I had a misguided brain wave. We carried pen flare guns as a bear deterrent. I thought, "I'll fire the flare gun over their heads and scare them."

Wedging my hand between the window and the protective screen, I pulled the trigger down and released the flare. The flare went shooting off but unfortunately didn't travel vertically, instead it flew obliquely towards the fuel tanks! I almost blew up the entire camp and ended up with the nickname 'Rocket Man.'

We were often hungover in the morning as we headed off to another scorching day in the muskeg. One morning I had stayed in camp to work on some plant identification. I was working in a portable work shelter when I heard a rustling noise in a tarp on the ground outside the shelter. I went to see what had disturbed me and was amazed to see Len Knapik rolled up in the tarp, fast asleep. Len, who was in no condition to work that day, had found a hiding place to sleep off his hangover.

The summer passed in a hot haze of forest fires, pot smoke, and alcohol fumes. It was amazing that no one was very seriously hurt while working or partying. None of us could have imagined how the oil sands developments and population in Fort McMurray would explode in the next 30 years. There was $150 billion poured into the region between 2000 and 2008; the regional population ballooned to 116,000 by 2015, with another 35,000 transient workers regularly ensconced in work camps. The blush is off the oil sands these days, but the city is still a bustling place with traffic problems even though it is 430 kilometres north of Edmonton.

The black flies were a constant annoyance while we worked. Black flies are part of a huge group of genera with more than 2,000 distinctive species. Along with mosquitoes, they are attracted to human CO2 emissions, dark clothing, and the scent of soap. When the fly bites, its saliva carries an anaesthetic and anticoagulant. Hence the fact that we would not feel the obnoxious critters biting us. They would often go for the area behind the ears, and we would find a stream of blood running down our necks. They loved to get under our clothing and gnaw on our tummies, wrists, and legs. Often, we were covered with angry, red welts. I remember one day Larry lifting his shirt to reveal a string of bites around his waist. We applied endless applications of musk oil with a DEET [diethyltoluamide] content of 90% to our skin, but after about two weeks, our skin had become leathery. The chemical would melt pens and anything plastic. DEET was developed in 1944 and was initially used in the jungles during World War II. Today, DEET levels are rarely above 10% in modern insect repellents. Alarmed by its aftereffects resulting in leathery skin, we resorted to wearing mosquito netting to protect our heads. Craig, Larry, and I named ourselves after the Knights of Ni found in Monty Python's Holy Grail movie. In the movie, the knights demanded a sacrifice from King Arthur and were just silly with him. Like those knights, we too became silly, prancing around in the bogs wearing our mesh hats as the flies harassed us mercilessly.

11

THE GRAND CANYON OF THE STIKINE AND A JAMMED SHOTGUN

In 1982 Greg Utzig, a plant ecologist from Nelson, BC, Helen Habgood, and I took the ferry from Port Hardy at the north end of Vancouver Island to Prince Rupert on the mainland and then made the 725 kilometre drive to Dease Lake in northwestern BC. Our task was to conduct a broad-scale biophysical survey over the entire Stikine River drainage and prepare a report addressing the resource values. Pedology Consultants had been retained by the Tahltan First Nation in response to BC Hydro's plans to build a hydroelectric project on the Stikine River. The Stikine, over 500 kilometres long, begins its journey in the highest elevations of the Spatsizi Plateau, exiting into the Pacific Ocean near Wrangell, Alaska. The drainage is 52,000 square kilometres, and located about two-thirds down its length, is the stunning Grand Canyon of the Stikine, a wild, water-cut gorge 75 kilometres long and over 300 metres deep. The canyon is home to numerous mountain goats that scale the cliffs secure from predators. Mountain goats are among the most proficient climbers in the ungulate family, moving easily up and down near-vertical cliffs. The watercourse is considered unnavigable, although hardcore kayakers consider it the 'Mt. Everest' of whitewater boating.

Our helicopter pilot for the project was Paul, a young daredevil Viet Nam war vet. He liked to fly fast and just over the treetops.

We had to train him to fly the way we needed, which was relatively high above the canopy, to be able to observe the lay of the landscape and accurately navigate our route during the day. Over the course of my career, I came to realize that older pilots were more reliable and trustworthy than the young bucks who liked to take chances and impress the hapless passengers. Paul was one of those young chopper jockeys. During one of our flying days, he offered to fly us through the canyon. Greg and I were gung-ho, but Helen, who was a nervous flyer, was not too sure. The top speed of the JetRanger is 222 kilometres per hour, and we screamed into the canyon at the eastern end about halfway below its height. The trip thru the canyon, twisting and turning, only lasted about 20 minutes, but what a thrill ride it was. As we angled our way through the sinuous route, we saw countless goats perched on the vertical cliff walls and seething, boiling rapids below us.

We completed a week of fieldwork out of Dease Lake, then drove 108 kilometres to the west and Telegraph Creek, located on the banks of the Stikine River below the canyon. Telegraph Creek is a small town of approximately 250 people, predominantly Tahltan residents. The town is a relic of the 1860 gold rush and takes its name from the Russian-American telegraph line. The line, an ambitious undertaking, was never completed and was abandoned in 1867.

Our lodgings were a ramshackle set of cabins owned and operated by the Tahltans. At the end of one workday, upon returning to our humble cabin, I was stunned to see that my bottle of Scotch whiskey was missing and had been replaced by a large sockeye salmon. Obviously, one of our Tahltan landlords considered it to be a fair trade, and although I missed the elixir, I had no problem with accepting its replacement.

Ultimately, with continued pressure from the Tahltan Nation and conservation groups, BC Hydro abandoned the project plans. The Grand Canyon of the Stikine was forever protected in the establishment of the Stikine River Park in 2001.

Later in the summer, Denny Maynard, an accomplished terrain mapper, and I were tasked with collecting vegetation and soil data in the Tahsish River and Kois River drainages on the northwest coast of Vancouver Island. Denny and I caught a ride with our helicopter pilot, Pete Barrett, in Port McNeil. We loaded up our gear, including tent, sleeping bags, and other camping supplies as well as food. We also had with us a shotgun which was given to us by our boss, Mark Walmsley, one of the owners of our company, Pedology Consultants. Pete deposited us near the mouth of the Tahsish River, and as darkness settled, we went about putting up camp. I wanted to test out the shotgun, which we had with us as protection from bears. I loaded slugs into the chamber, but as I did this, one of the slugs got jammed, and I couldn't extricate it. Now, we had a gun but no way of using it. We hoped that we would not encounter any obnoxious bears. The next morning, we developed a strategy for sampling the valley and quickly realized that there was no way to get to the other side of the river, where camp was. The only way to get there was to wade across the river buck naked with our clothes, boots, and field gear held above our heads. We were able to accomplish this without being swept downstream.

We accessed most of our sample sites by hiking to them, and the valley was a fascinating unspoiled environment without any logging and full of old-growth forests. The valley was home to a large herd of resident Roosevelt elk. On one of our hikes into the upper part of the valley, we came across a large group of the animals. The elk are magnificent ungulates (herbivorous mammals with hooves) with a combination of pale and dark brown areas on their bodies; bulls can be up to 350 kilograms, with the largest ones reaching more than 500 kilograms. The huge antlers of bull elk make it one of the most desirable trophy animals for hunters from all over the world. It is not uncommon for a mature bull to have antlers more than 125 centimetres long. The antlers are shed each spring and grow back every year. At a growth rate of up to two and a half centimetres per day, it is some of the fastest-growing tissue in nature.

After a week of collecting ecosystem data, we had Pete fly in and move us to the Kois River floodplain located near the head of the Tahsish River. When he came to pick us up, he brought Terry Vold, a government soil scientist, with him to join us for two days of final sampling. We set up our tent on a gravel bar in the river, and that night, as the three of us were wedged in our small two-man tent, we were subjected to the horrendous snoring of Terry which made our sleep challenging, if not impossible.

The next morning, we awoke to a violent wind and rainstorm. Within hours the gently flowing Kois River had been turned into a raging torrent, and we quickly had to move our tent and camp to a higher terrace to prevent us from being swept away. The worst part was that the beer that Terry had brought was lost in the torrent. It was raining so hard that we barely ventured out of the tent for two days. Ultimately, Pete came back to extricate us, and we never obtained much data in the upper Kois due to the storm. In 1995 the BC provincial government established the Tahshish-Kois Provincial Park, ensuring that the area would remain untouched in perpetuity.

SECTION 3

Working Independently

I had little choice but to start my own business when the recession hit in 1983. With no business plan, I learned as I went. I was lucky that my years in the civil service had left me with a ready-made list of contacts that helped me find enough work for the next seven years. I had always been independent, so had no trouble adapting to working at home. It also gave me an opportunity to be around during the time when our children were growing up.

12

SURVIVING THE RECESSION AND DAY OF THE GRIZZLY

In 1983 Canada was in the middle of a crippling recession, and I became a job loss statistic when the work at Pedology Consultants dried up. With no work options, no business plan, and the future looking gloomy, I started my own proprietorship called Shearwater Ecological Services to provide consulting services to the government and private industry. Pedology had dwindled to a few staff, and those of us remaining were on a federally funded work-sharing program. Without many avenues open to me, going independent was an easy choice.

In 1984 I was contracted by the MOF (BC Ministry of Forests) to conduct an ecosystem inventory of the upland portion of the Kimsquit River drainage. The Kimsquit flows into Dean Channel approximately 200 kilometres north of Bella Coola on the mid-coast of BC. The MOF staff, including Jim Pojar and Allen Banner, both exceptional plant ecologists, were responsible for the river valley portion (floodplain) of the study area while I sampled the sides of the valley. I needed a field assistant, and I hired John Klarer to help me. John was the young, keen son of my landlord and had no field experience. I figured youthful experience would trump lack of bush savvy. I was wrong.

The ecosystem mapping was being done as an input into the development of a management plan for grizzly and black bears. The lead bear biologist was Tony Hamilton, a lean, wiry man with a shock of dark, curly hair and an easy laugh. From Tony, I learned how to be safe in bear country: always keep your eyes and ears open, make lots and lots of noise, and watch for signs including scat, mark trees, and mark trails. To let the bears know we were coming, we would yell, "ANY BEARRRRRRR!" as loud as we could. If we saw scat that was shiny or had flies on it, then it was a sure thing that the poop had been deposited recently. If it was steaming, then it was time to leave immediately. Mark trees were typically young evergreen saplings. Bears would claw them to expose sap, then rub up against them to leave scent and hair. The scent and hair were left to notify other bears of their territorial claim. Mark trails were the most interesting phenomena. The trails appeared most often on floodplain sites where grizzlies would use the same route habitually, leaving giant footprints in the ground. Years later, Tony would capture video evidence, with remote trail cameras, of bears placing their feet in the tracks and twisting their limbs to increase the depth of the impressions. Over the years, I saw many of these mark trails in bear country, not only for grizzlies but also for black bears.

During our two-week field trip, we stayed at a logging camp near the mouth of the Kimsquit River. We slept in camp rooms, and the whole team ate meals, which we prepared ourselves, in a large log cabin.

The budget for the project was minimal, with limited funds for helicopter use. Consequently, John and I would use a beat-up old truck to drive up the valley on a daily basis and then hike about 1,000 metres of vertical elevation from the road up into the subalpine and then trundle back down at the end of the day. The slopes were steep with thick brush. Lower elevation sites were a mixture of western hemlock and western redcedar with large Douglas-fir on warm sites with shallow soils over bedrock. The subalpine forests were open with well-spaced mountain hemlock and yellow cedar; the shrubs, copper bush and Alaskan blueberry formed an oft impenetrable layer. On our way up, we completed ecosystem plots at selected locations. John's job was to dig soil pits and age trees using an increment borer. I would compile a plant species list and estimate cover ratings. The daily grind of hiking uphill was tough slogging, but we sure got in good shape. John grumbled a lot, especially on rainy days when the footing was dicey. What often took six to eight hours going up, we would cover in about an hour going down.

On one occasion, John and I were trekking downhill and ended up in a steep, rocky area with some portions too dangerous to traverse. Even though we used aerial photos to plot our routes, it was impossible to pick out rocky bluffs. We were finding our way along an area with steep bedrock sections and gullies when I suddenly saw two black bear cubs scoot up a tree below us. My immediate reaction was, "Aw, isn't that cute!" which was soon replaced by, "Uh oh, where is momma?" I figured if I chucked a boulder down the gully, we could locate her and avoid contact. I quickly grabbed a large rock and heaved it down the hill. No sooner had I done that than we immediately knew where momma bear was. She was charging up the gully towards us and in a very ornery mood. Her fur was all erect, ears flattened, and she was huffing at us. As she got closer to us, my primitive 'fight or flight' instinct took over. Flight was the only opti0n, and adrenaline kicked in as my legs worked like pistons. I ran across the rock slope like a deer. "Flee," was all I could think. I looked back and saw that John was close behind me. Momma was nowhere in sight, and we kept going until we reached the safety of our truck at the road. Only then did I realize that John didn't have the backpack with our field equipment with him. He sheepishly admitted that he had dropped it in his panic when the bear charged us. The next day we retraced our steps back up the hill and luckily were able to find our gear. I was amazed when I saw the terrain that we ran across to escape the bear. Adrenaline is a powerful stimulant in times of crisis, and I experienced it a few times in my career. I had heard stories of people able to leap high into trees when being attacked by bears.

Another day, John and I had hiked partway up a steep slope and were doing a plot. As we worked, we heard the sound of people approaching us from below. It was Jim and Allen, unaware of our presence. We played a mean trick on them. As they got closer, I made a deep growling sound, trying to imitate a bear. They both started shaking their bear bells vigorously, and I am sure they were a trifle scared. John and I were laughing as we revealed ourselves to them, and they both looked relieved.

As part of Tony's grizzly research program, many bears had been fitted with radio collars so they could be tracked with telemetry equipment. Typically, his crew would monitor a bear's movements, and then the day after a bear had shifted locations, they would search the activity area to document habitat use (especially focusing on the plant species being eaten).

On one such occasion Andy Derocher, wildlife biologist and Scott Lindburgh, an MOF soil scientist, ventured up the valley to check on the activity of grizzly bear number 823. Telemetry showed that the bear, a large male (boar), had moved overnight. They approached the habitat use location but had a chilling encounter with the bear that was much too close for comfort. The telemetry had been horribly wrong. As they neared the site in question, they heard the massive animal charging towards them. They couldn't see the beast in the thick undergrowth, but the thrashing movement of the shrubs showed that the bear was coming at them like a runaway train. Top speed for a bear is about ten metres per second. Andy was armed with a shotgun loaded with slugs and raised it to his shoulder as the bear suddenly emerged from the brush in front of them, in a foul mood and roaring at them. He fired a shell over the bear's head and watched as the animal veered away to the side. It was a bluff charge!

When Andy and Scott returned to camp, they were shaken and pale, severely shocked by the encounter (and a few years older). Neither one of them ate dinner that night. A year later, I would have my own encounter with a nasty bear.

13 BEAR SCARE AND MAN OVERBOARD

The Bell-Irving wildlife habitat project, in 1985, involved ecosystem mapping in the Bell-Irving River drainage in northwestern BC. I worked with Larry Lacelle, a soil scientist and Grant Hazelwood, a consulting wildlife biologist, on the job which was for assessing ungulate (moose and mountain goat) and bear (grizzly and black bear) habitat. Larry and I drove from Prince George to Meziadin Junction where there was a run-down camp, a dilapidated gas station, and a tacky restaurant. It was on Highway 37, about 153 kilometres north of Kitwanga on the Yellowhead Highway (Highway 16). The camp was relatively primitive, and a bunch of battered car and truck wrecks behind the gas station attested to the dangers associated with driving on the rough road that headed north to Cassiar and the Yukon. All our sample sites were accessible only by helicopter. Our pilot was Gary, a happy-go lucky speed merchant with a penchant for chasing wildlife, which is strictly illegal.

One bright, sunny morning, Gary dropped us off in a small wetland near an old-growth (greater than 200 years old) western hemlock forest. He said he had to go for fuel and would be back in about half an hour. That would give us time to do our plot. We walked about 100 metres into the forest and began to work. Larry was digging his soil pit, Grant was walking around looking for wildlife signs, and I was starting my list of plant species.

Suddenly we heard a loud crack, and there, no more than 50 metres away, was a large black bear leaning over a horizontal deadfall. Oh shit! He was acting aggressively, ears flat, growling at us, and snapping his jaws together. We had no weapon, save for Larry's shovel, and so tried scaring him off by yelling at him and banging metal objects together, to no avail. He was prowling towards us, circling around us, coming to within only 20 metres away, and then a mere 15 metres from us. Grant quickly told us that we should go back to the wetland as black bears generally don't like being out in the open. There was no sign of Gary returning yet. We hastily, and as calmly as we could, walked back to the wetland where we stood close together with our arms in the air to make ourselves look as large as possible. The bear followed us out to the edge of the fen, where he started smashing small trees down and continued to growl loudly. Still no Gary. We spent a nervous 15 minutes watching the bear as he continued his aggressive behaviour.

Finally, after what seemed like an eternity, we heard the unmistakable sound of the helicopter approaching, and Gary arrived to save us. Grant was convinced that the bear was a predatory male and, given the opportunity, would have made a meal of one of us. Interestingly, a year later, a rig hand working alone was killed and partially eaten not far from our location. In my mind, it was likely the same bear that had confronted us.

Right after the bear scare, we flew up into the alpine, where we thought we would be 'safe' from marauding bears. Grant flew off with Gary to look for critters as Larry and I were doing a plot on a grassy slope. No sooner had Gary dropped out of sight in the machine than we heard the chopper coming back towards us. As we looked up, a small herd of about ten goats came charging up the hill towards us, being steered towards us by Gary. I could see him laughing as the goats sped by us with their tongues hanging out and looking panic-stricken.

When I got home after the trip, I told my wife, Anne, I would acquire a shotgun. The experience with the bear had shaken me badly, so much so that I figured it was time to get some form of protection while venturing into the wilderness.

My neighbour Don, a hunter, made a sawed-off shotgun for me by cutting the stock back about 15 cm (centimetres) on a normal shotgun. The first time I fired it, I wedged the stock against my hip, and it damned near broke it with the recoil. I had a massive contusion and sore spot for weeks. I soon realized that the only way the gun could be fired was to hold it to the side of my body, allowing my arms to take the shock. The gun could hold five slugs, and for ten years, I carried it religiously whenever I was working in bear habitat.

Eventually, I came to the realization that in a bear attack, I would probably not fare well even though I was packing the 'heat.' A bear in a full-out charge would cover about 10 metres per second and given that I would not be able to fire a slug until he was within 20 metres, it dawned on me that I would likely have only a scant two seconds in which to fire one slug certainly, but two slugs not likely. Another factor would be that in most cases, the bush would obscure my vision of any bear charging towards me. Plus, a bear charging at me would present a small target, essentially a round cylinder and not a large standing target.

In the ten years of packing the gun, I only ever had cause to fire it twice. The first time I was with my son Colin working near Pemberton, BC, when we came across a black bear sow with a cub. My warning shot fired over her head was enough to motivate her to move away from us. The second time, I was following a game trail along a rushing creek on a rainy day near Bella Coola, BC. I came around a corner and saw a large brown shape, likely a grizzly bear, in front of me. Not taking a chance, I fired a slug over the bear's head as it ran away from me. In that incident, I was convinced that the bear had heard my warning bell attached to my backpack before we were even close to it.

In the fall, Dave Yole, an independent soil scientist, and I were engaged by the MOF in Smithers to conduct biogeoclimatic classification sampling along the sides of Observatory Inlet on the coast of BC by boat. Biogeoclimatic sampling included collecting vegetation and soil information at selected sites to allow for identification of ecosystems. We met up in Smithers and then made the long drive to Stewart, BC, about 330 kilometres away. Stewart is located at the head of the Portland Canal, which is about 114 kilometres long and forms the boundary between southeastern Alaska and BC. Stewart was a boomtown prior to World War I when gold and silver mines blossomed in the region. The population then was close to 10,000. Its next-door neighbour, Hyder, Alaska, is right across the border. In Hyder, a famous tradition exists, which involves downing a shot of 150 proof Everclear alcohol in the Glacier Inn. As you could expect, Dave and I did, in fact, get 'Hyderized.' In 2011, the populations of Stewart and Hyder were respectively 495 and 87.

Once in Stewart, Dave and I met up with our captain, Ralph, and his boat, the MOF Coast Ranger. The Coast Ranger was a 15-metre patrol ship built by the BC Forest Maintenance Depot in Vancouver in 1967. We loaded up our field gear and headed down the Portland Canal towards Observatory Inlet, which ran northeast from the canal about halfway down its length.

On the way down the canal, we were greeted by a group of Dall's porpoises riding our bow waves. The Dall's porpoise is a speed demon and can attain speeds up to 55 kilometre an hour or more. The species is well known to boaters on the BC coast for its habit of riding the bow waves of passing boats. They look like a little black and white torpedo. Fully grown, they are only slightly more than two metres long but stocky and powerfully built, weighing up to 220 kilograms. The porpoises stayed with us for quite a while, and then as suddenly as they had appeared, they disappeared. Eventually, we arrived at the intersection with Observatory Inlet and headed north to begin our survey. We would pick locations where the vegetation looked suitable and then have Ralph run us into shore in a small motorized dingy, which we were pulling behind the Coast Ranger. We had a hand-held radio with us, and when we were finished with our description of vegetation and soil, we would call Ralph to come and retrieve us.

One day when we called Ralph on the radio, we did not receive a reply. We tried calling him repeatedly with no luck, and after numerous attempts to raise him, we became concerned. We wondered what could have happened and contemplated having to swim out to the ship. After about half an hour, we finally saw Ralph approaching us in the small boat. When he reached us, we could tell something bad had happened. It turned out that after he had dropped us off and took the boat back to the Coast Ranger, he had slipped off the deck upon his return and fallen into the water. He had struggled to get back onto the mother ship and eventually was able to haul himself back on deck. He had nearly drowned and, in the process, lost his glasses, almost becoming hypothermic in the cold water. That incident caused much alarm with his supervisors since he should not have been managing the vessel on his own anyways. As soon as they could, the ministry sent a mate to join us and ensure that such an accident would never happen again. It was our good fortune that Ralph had run a restaurant in Tofino during a past life. He was an excellent chef, and we had many fine meals during our time with him.

Halfway up Observatory Inlet, we encountered the abandoned town of Anyox. It had been another one of BC's boom towns in the early 1900s driven by mining of silver, gold, and copper but now lay deserted and abandoned. The coast of BC has a litany of ghost towns similar to Anyox that were typical of the 'boom and bust' cycle of resource extraction. Now, as we cruised past its abandoned ruins, imagining its glory days, we felt a sense of desolation[3] (see Footnotes and Appendix for more information on Anyox).

The rest of our fieldwork passed uneventfully, and we were sad to say goodbye to Ralph and the Coast Ranger when we finally disembarked back in Stewart. The trip with the ship used to access our sample sites ended up being one of the most unique in my career.

That year I became an RPF (Registered Professional Forester) in BC. While most professional foresters get a Forestry degree at UBC before writing the policy exam, I never went to university. I had taken another route, which consisted of completing ten correspondence courses through the Guided Independent Studies program administered by UBC. It was a tough row to hoe. Over a five-year period, I completed the course requirements and then barely passed the policy exam.

14 THE MUM FIRE AND FIELD TRIP WITH KOJI

In the summer of 1986, I had a contract with the MOF in Smithers to assist with the biogeoclimatic sampling program in the Entiako area, located about 200 kilometres south of Burns Lake in central BC. We did two field sessions: the first one was working out of a camp which we had to set up ourselves at the west end of Johny Lake, and the second one involved working out of cabins near Moose Lake. My other crew members were Jim Pojar, Allen Banner, Andy MacKinnon, a MOF ecologist from Victoria, and Doug Holmes, a soil scientist. Part of the team on the first field trip was Karen Price, a young biology student volunteering her time. Doug was one of the most hirsute men I had ever met, with his entire body covered with a profuse layer of hair. Without his clothes on, he looked like a huge teddy bear.

For the first trip, we used a helicopter and a Zodiac to access our sample sites. One day, Jim and Andy had used the zodiac to cross Johny Lake then run a transect through the bush to collect vegetation and soil information along their route. Jim prided himself on being able to navigate through the bush without the aid of a compass. At the end of the day, when they returned to the lake to find the Zodiac, they were shocked to find that the rubber raft had 'disappeared!' They were flummoxed. Only after much thinking and looking at their aerial photos did they come to the realization that Jim's navigating had brought them to the wrong location. They had to backtrack and, after much tramping through the woods, were able to finally find the correct waterbody and retrieve their mode of travel across the lake. When they finally arrived back at camp, a couple of hours late, we all derided them for their mistake. It was an error in judgement that Jim never made again.

One day when Karen and I were working together, we had been dropped off by the helicopter to complete two ecosystem plots. We were in a young lodgepole pine stand where I was busy taking notes as Karen had her head down in the soil pit, looking for the soil horizons that indicated the type of soil development. Suddenly I saw a large brown shape running through the trees directly towards us. It was probably a young grizzly coming over to investigate the sounds he had heard. Without a weapon and no bear spray, my primitive man instincts took over. I immediately grabbed a large branch and started whacking it against trees as I yelled loudly. I was relieved to see the bear quickly turn and run away from us. Karen, with her head in the pit, never saw the beast, and she had no idea why I had acted so strangely. When I excitedly told her what had happened in those few seconds the encounter had lasted, she looked scared and relieved at the same time.

One evening after we had all eaten our evening meal and were relaxing around the campfire, it was my turn to wash the dishes. This would consist of gathering up all the plates, cutlery, and pots and pans and washing them at the edge of the lake, about 15 metres away from camp. I was busy cleaning the dishes when I heard a crack off to my left. I was surprised to see flames and smoke coming from the vicinity of our latrine site. I quickly ran back to the campfire and hurriedly told everyone we had a fire on our hands! It had been a dry summer, and the fire hazard rating was high; a forest fire was the last thing we needed. The whole crew, including our helicopter pilot, ran to the latrine site to see an area of about 10 metres wide engulfed in flames. We sprang into action, using buckets, pots, anything else that would hold water, as well as picks and shovels to put the fire out. After about an hour, we were successful in extinguishing the flames and bringing the fire under control. During our frenetic labour, I had noticed that one member of our crew was working more feverishly than anyone else. It was Andy. I was pretty sure that the fire had been caused by him smoking a joint while over at the latrine, with the 'cherry' (the glowing, hot end of the joint) falling out that had smoldered for a while before starting the fire. Andy never admitted it, and I never confronted him, but I am pretty sure that was what happened. We never told anyone else about the fire and we were lucky that it was never reported to the MOF. We always referred to the fire as the 'Mum Fire' since our lips were sealed. Hence, we applied the expression 'mum's the word' to our close encounter with disaster.

Later in the summer, Jim, Allen, Dave, and I were dropped by helicopter at the west end of Moose Lake where there was a small fishing lodge. We stayed in a large cabin with four bunks and were able to use motorcycles and three-wheeled trikes to access our sampling sites. The motorbikes were a breeze to handle, but the three-wheeled trikes were beasts. If you weren't careful, the rear wheels were prone to ride up the back of your leg and pull you off the machine.

Allen and Jim were fond of their 'Trapper Nelson' backpacks which were large and hefty but made to carry a large amount of material. The Trapper Nelson backpack was invented by Lloyd Nelson, who, in 1920, when struggling with the discomfort of a makeshift pack frame made from willow and sealskin, devised a different way to carry heavy loads in relative comfort. One night, Jim and Dave played a mean trick on Allen by placing a small rock in the bottom of his backpack. The rock was deep in the bottom of the pack where Allen wouldn't notice it, and sure enough, he went through the next day of fieldwork without finding it. Jim and Dave repeated the same process the next evening, adding another small weight to his pack. Finally, after about five days of this, we were sitting around the cabin in the evening, and Allen complained about how heavy his pack was. Only then did the pranksters reveal what they had done. Of course, Allen was furious, but Jim, Dave, and I all had a good laugh at his expense.

In the early fall, Jim gave me another small contract to collect site series (ecosystem) field data in the rugged, northwestern corner of BC. As usual, I had no regular field assistant. I don't recall how it came to be, but Anne's brother John (Koji) ended up flying out to BC from Montreal to join me on the trip. We flew to Prince George where we rented a 4 x 4 truck, then headed off to Meziadin Junction. John had no 'bush' experience but was a keen participant in another one of my field adventures. Almost the entire study area was festooned with vigorous devil's club, and because it was autumn, the leaves had fallen from the plants, leaving the bare stems covered in the nasty spines. We would often step on them and as we took another step, moving our boots and thus releasing the stem, which would lash us, sometimes in the face. We were heading into bear country, and I now had my shotgun to protect us.

We had a few helicopter days, and our pilot was Gary, the same fellow as the year before. John enjoyed the thrill of ripping around in the mountains, and one day as we approached a wide valley, we flew over a massive grizzly. Gary wheeled the machine around in a tight turn to get another look at the impressive specimen. As we came up behind the bear, who was now running at full speed across the floodplain, he turned his head to see who his pursuer was. I still remember the look on the bear's face and a display of huge teeth as he roared his displeasure. I told Gary to keep going as I wanted no part of his potential revenge.

On another occasion, John and I had landed in a wetland and walked about 30 metres away from the chopper into an old-growth western redcedar forest where John was digging a soil pit. Our radio came to life. It was Gary. "There's a grizzly out here," he said in a quiet voice, "he's walking past the chopper." We held our breath, and I took the safety off the shotgun as we stayed crouched and quiet. Luckily, the bear kept going and slowly moved away from us and the helicopter.

The rest of the trip passed without incident, and John left for Montreal with some amazing experiences and a good set of calluses on his hands. I know John had fond memories of that trip for the rest of his life. Sadly, he passed away in March 2020.

15 LOGGING CAMPS AND BEAR SCAT

In 1987 I joined with Tony Hamilton and his assistant Coleen Bryden, now a Senior Project Manager with Stantec, to start work on an interesting research project. Tony is retired now from the BC government, but back in those days, he was the main guy for grizzly bear work on the BC coast. The project was funded by FRDA, a joint BC-Canada Forest Resource Development Agreement which poured $300 million into the BC forest industry between 1985 and 1990. The survey was designed to assess the impacts of glyphosate, sold under the trade name of Roundup, on coastal grizzly bear habitat. Glyphosate was the main herbicide used by forest companies to control the vigorous shrubs on productive, logged forest sites that made establishment of new evergreen trees difficult. Tony had established in the early 80s that grizzlies spent more than 60% of their time on the most productive sites (both for forestry and bears), which were floodplains and fans. Fans are caused by the flow of water down gullies where relatively fine soil material is deposited at the base of slopes.

Over a two-week period in the summer, we travelled to four different coastal drainages to collect data on the vegetation and soil of selected sites that we knew were being heavily used by grizzly bears and were also being treated with glyphosate (we repeated the data collection at the same locations in 1988). The four drainages were the Chuckwalla and Kilbella Rivers (flowing into Rivers Inlet), the Kingcome River (flowing into Kingcome Inlet), the Klinaklini River (flowing into Knight Inlet), and the Phillips River (flowing into Phillips Arm). All these rivers are massive watercourses reaching the Pacific from the towering Coast Mountains. All of them are home to extensive populations of grizzly bears, and they are all wild, rugged, and untamed valleys save for a long history of logging.

Logging in these valleys had been going on since the early part of the 20th century. The logging camps where we stayed while we did our work were throwbacks to a bygone time and place. The camps usually consisted of a few dorm trailers, a dining room, and a recreation hall. Sometimes there was no recreation hall, and there were never TVs in individual rooms in those days. Additionally, there was usually an office, a first aid room, and the mudroom. The mudroom was where workers would hang their outer work clothes and caulk boots at the end of the day; there were heaters that were always cranked up to ensure that everything was dry for the next round of fighting the elements. It rained a lot on the coast, even in the summer, so the mudroom was always full of wet clothes and boots. The room had its own special mystique and attraction. When you were in the room, it was comfortable and warm, like being immersed in a steamy sauna. There were different odours wafting through the air. We would smell the needles of evergreen trees such as western hemlock, Sitka spruce, and western redcedar, there was sawdust scattered on the floor, which exuded a smell of freshly cut wood, and there was the heavy scent of dubbin, a water-resistant paste that was applied to leather boots. The camps were hewn from the rough ground and seething vegetation around them. There were always decaying, rambling, ramshackle buildings in the vicinity of camps, reminders of the times when men would live at these temporary accommodations for months on end. The old buildings were being invaded by red alder, salmonberry, willow, and thimbleberry trying to consume them. The camps were full of rough, sinewy, and resilient men. In those days, you never saw any women at the camps. You had to be tough and ornery to survive working in those conditions. There was always a pecking order present, with the fallers, who had the most dangerous job of all, working on steep slopes to cut down the forest giants at the top of the food chain and the chokermen at the bottom. The chokermen were the poor souls who scrambled around in the logged settings, placing cables around the felled trees so they could be hauled up the slope to the logging landing; at the landing, the trees had branches removed and then they were loaded onto logging trucks. While I was at BCIT, I had applied for a summer job as a chokerman. In the interview, I was told by the company rep that I was too big to be a chokerman (I am over six feet tall). They liked the chokermen to be small fellows so they could easily dart under and around the massive mess of logs on the ground.

We were outsiders at the camps, and the regular inhabitants were always curious about why we were there and what we were doing. As soon as we told them that we were working with grizzlies, their respect for us went up a notch. We never bothered to relate any details about our actual research. Just the mention of grizzly bears was enough to impress them.

At one camp, poor Chip (Colleen's nickname because she had a penchant for chocolate chip cookies) was set upon by two drunk loggers. The concept of a 'dry' camp didn't exist back then. Chip was in a separate dorm, and these two louts decided that they wanted to get friendly with her. They stood outside her room, talking sweetly and trying to entice her out. Terrified, she refused to budge and stayed huddled inside while they wallowed about in the hallway. She didn't tell Tony and me about the incident until the next morning. We talked to the camp manager about it, but there was no way of tracking down the perpetrators. We told Chip to get us the next time it happened and we would set them straight.

Chip also had a tough time with no-see-ums. No-see-ums are tiny flies (they are also called midges, are almost invisible, and are less than one millimetre long), and, like mosquitoes, the females need blood to reproduce. There are about four thousand species of this type of insect, and they are pests capable of giving a nasty bite, despite their small size. Clouds of them would come out in the cool morning air, then just as quickly, when the sun came up, and air temperatures rose, they would disappear. Chip didn't wear gloves when we were working, and consequently, her hands would be covered with many bites. She would be scratching them incessantly over breakfast, and every day I would admonish her and tell her to stop. If you scratched the bites hard enough, skin infections were sure to arise.

Daily, we would drive down active logging roads, dodging loaded, off-road logging trucks on our way. Off-road logging trucks were behemoths, up to eight metres wide, and if there was one coming at you on a blind corner, carrying over 50 tonnes of trees, there was nowhere to go but in the ditch. Driving a logging truck down steep, rough logging roads was, and still is, the most dangerous job in the logging industry. In 2005, BC recorded 41 deaths in the forest industry, and many of those were logging truck drivers[4] (see Footnotes for source of information on logging industry fatalities).

At some point on a road, near where we could see suitable logged, floodplain bear habitat, we would park our truck and then walk into a sample site. Once at the site, we would record data on the plant species present, soil conditions, bear signs, and the effects of glyphosate application. We would regularly see grizzlies in the field, but never close enough for us to be concerned. We were at one such site, with dense and continuous cover of salmonberry (more than one metre tall), recording data when we had a bear encounter. Suddenly, no more than 10 metres away, the shrub cover started moving vigorously. Something, probably a grizzly, had woken up and was stirring. We yelled loudly as the bear continued to move without us being able to see him or her. Our heart rates were quickly elevated, and there was nowhere to hide if the bear decided to attack. Fortunately, the bear continued to move away from us, and we never did see its countenance.

On our trips back to camp, Tony would stop the truck often to collect samples of bear scat. Bears would regularly leave poop piles as they walked down logging roads. Tony took the scat samples back to camp, and in the evening, he would weigh the samples and take them apart to try and figure out what the bears had been eating. He would often call me to help identify the plant species that had been on the menu from looking at the seeds in the scat. Not a particularly enjoyable task but one that was part of the research requirements.

Occasionally we would come across a large tree stump in the forest that bore signs of an ancient logging practice. In the late 1800s and early 1900s, loggers, called lumberjacks back in those days, would work in teams of two or more workers to fall massive trees that were typically more than four or five metres in diameter. It was a monumental task to harvest these giants. They would cut notches at the base of a tree, insert 'springboards' into the trunk, and stand on these boards wielding a long, crosscut saw up to three metres or more in length. It would take hours to bring the tree to the ground. Then they would buck the tree up on-site or use teams of oxen to haul the tree to a river, lake, or the tidewater, where it would be transported to the nearest mill. With the advent of the mechanical chain saw in the 1940s, the hand loggers became a dying breed.

The practice of hand logging had its origins in the 1860s when the first sawmills were built around the Gulf of Georgia. Hand logging started in Burrard Inlet, near Vancouver, to supply the mills with a steady source of wood and then spread up the coast. From simple beginnings, hand logging became akin to the development of a gold rush in slow motion, and many Cariboo and Klondike miners found their way into the ranks of the hand loggers when the mines played out.

16

THE CHILCOTIN CARIBOU AND THE GREAT BEAR RAINFOREST

In the summer of 1988, I had a contract with the MOE (BC Ministry of Environment), Wildlife Branch in Williams Lake, to conduct a woodland caribou habitat mapping project in the Chilcotin area. The Chilcotin is a vast swath of dry lodgepole pine forests (primarily on coarse-textured sand and gravel deposits) located west of Williams Lake and north of Highway 20. The highway runs from Williams Lake to Bella Coola on the BC mid-coast for approximately 460 kilometres. My field assistant for this project was Mike Robinson, a teacher who although available in the summer, had limited field experience. It would turn out to be an illuminating experience for him. Mike and I had met while playing recreational volleyball in 1985 and ended up becoming close friends. To this day we have a long-term golfing relationship that has endured for more than 35 years.

The mountain pine beetle emerged as a threat in the late 1990s, endangering all of the lodgepole pine forests in the interior of BC, including those in the Chilcotin. The beetle, which is naturally occurring, had been kept in check prior to the 1990s by typically cold winters that killed many of the beetle larvae that overwinter beneath the bark of pine trees. With the advent of climate change and resultant warmer winters, there was no naturally occurring mechanism for control and the populations exploded. Where Mike and I were working, there was not yet any evidence of mountain pine beetle outbreaks.

Our focus was on the pine forests where terrestrial lichens (growing on the ground – lichens that grow on trees are called arboreal lichens and include members of the *Alectoria* and *Bryoria* genera) flourished. They provided a winter food source for possibly the largest caribou herd in southern BC. The caribou focused their feeding primarily on three members of the *Cladina* genus. The three most important species were: *Cladina rangiferina* (grey reindeer lichen), *C. stellaris* (northern reindeer lichen), and *C. mitis* (yellow reindeer lichen). Lichens, in general, are amongst the most primitive and curious of plants. They are not a single species but are comprised of a fungus and algae, which are symbiotic. Being symbiotic, the basis of their relationship is the mutual benefit that they give to each other. The photosynthetic algae, or cyanobacteria, form simple carbohydrates that, when released, are absorbed by fungi cells and transformed into a different carbohydrate. The algae also produce vitamins that the fungi need. Fungi contribute to the relationship by absorbing water droplets from the air and by providing shade for the light-sensitive algae beneath.

Lichens are the principal winter food of caribou and other ungulates. Although hoofed mammals can survive for several months on a lichen diet, they usually lose body weight because of protein deficiency. The complex carbohydrates of the lichens are broken down by enzymes produced in the stomach of caribou and reindeer (reindeer are native to Scandinavia) but in few other animals.

Mike and I drove from Victoria to Williams Lake in a beat-up, old pickup truck, met the client representatives, and then continued on to Chezacut and Roy Mulvahill's ranch, about 43 kilometres north of Alexis Creek on Highway 20. There, we stayed in a ramshackle cabin, surrounded by herds of turkeys and cattle. The cabin was our home for a week. From Roy's ranch, we drove in different directions to collect vegetation, soil, and caribou habitat data. We also utilized a helicopter for a couple of days and got dropped off by floatplane high in the mountains near a small lake for another few days. When the helicopter came to the ranch, all the turkeys gathered in a panic-stricken mob, thinking that danger was imminent and not knowing which way presented an escape to safety.

The ranch was located south of the Itcha and Ilgachuz Ranges, one of the most unique set of mountains in BC. The ranges are examples of isolated shield volcanoes rising to 2,400 metres above sea level. They lie in the rain shadow of the Coast Mountains; as a result, they are relatively dry, with many unique ecosystems. The mountains have a variety of interesting volcanic landforms in varying hues of yellow, white, orange, and red. In 1995 the BC provincial government designated the wilderness area of 112,000 hectares as Itcha Ilgachuz Provincial Park. I didn't know it in 1988, but I would be back in these mountains in 1996 and 1997 to do more habitat mapping.

Mike was fun to work with simply because he was a bush rookie and was amazed at the conditions we worked in and what he had to do. He had never been in a helicopter or floatplane before. On our first field trips with the truck, Debbie Cichowski, a caribou biologist, joined us to educate us on the ways of the caribou.

Under the open lodgepole pine stands, on the coarsest soil material, the lichens flourished. When the weather was dry, the lichens were crunchy underfoot, and when it rained or was damp, they were like miniature rubbery, soft trees. Debbie showed us how to lay out caribou scat density plots, with lines running for 15 metres out in cardinal directions (north, east, south, and west). Along each line, we would look for piles of caribou droppings, small dark-coloured pellets about one centimetre in diameter with one end pinched off. We had to record the number of pellet piles along the line and then give the site a rating of low, moderate, or high in terms of caribou winter use. The Chilcotin is an area of low winter snowpack, and as such, the caribou can 'smell' or sense the lichens through the snow cover.

As with any of our field projects in the summer season, we had to contend with mosquitoes, and Mike was particularly paranoid about the small pests. At the first sign of a few, he would extricate his defensive equipment from his backpack. The item he used was called 'Uncle Keith's Bug Jacket.' I am not sure where he got it, but it was impregnated with a chemical that smelled disgusting and was supposed to ward off mosquitoes. I am not sure if it actually worked, but it made him feel better.

Mike didn't own any real 'field' clothes. I remember one day when it was raining and he was tramping through the wet brush without the benefit of rain gear, wearing sweatpants. By lunchtime, the sweatpants were saturated and working their way down towards his knees. He was soaked through, and come the end of the day, we had to build a giant campfire to dry him and his clothes. I still have a good chuckle thinking of how he looked like a drowned puppy trying to keep his sweatpants up.

The last segment of our fieldwork included a few days working from a tent camp high in the mountains of the Ilgachuz Range. We loaded up our gear into a small floatplane at Nimpo Lake and headed into the backcountry. We had no problem landing on a small lake.

After saying goodbye to our pilot and arranging for him to pick us up in a few days, we set about establishing our camp. An existing latrine spot had already been established by previous campers. It consisted of a toilet seat secured with stakes and rope between two small trees, and the poop was then allowed to roll downhill on a steep slope. It was quite a unique gravity-assisted waste disposal system, one which I had never seen and would never see again. Personal hygiene consisted of washing ourselves in the cold mountain lake, and that process was always bracing with the near-freezing temperatures of the water.

After three days of roaming in the high elevation pine forests and collecting more ecosystem and lichen data, our pilot returned to take us back to Nimpo Lake. Once we had jammed all of gear and ourselves into the small plane, the pilot made a somewhat disconcerting comment to the effect of, "I sure hope we can get out of here with all this weight." Mike and I held our collective breath as we taxied to take off and then barely cleared the treetops on our departure. Over the years, I had many sketchy flights in small planes where I am sure we were over the legal weight limits. Given a choice between flying in a helicopter and a small plane, I would take the helicopter every time.

In 1988 I became an R.P.Bio. (Registered Professional Biologist) certified to practice Biology in the province of British Columbia. The APB (Association of Professional Biologists) was a neophyte organization formed to regulate the practice of biology in BC. I had been in the business of doing ecosystem fieldwork for more than 15 years, and the leaders of the organization asked me if I would be willing to join. Eventually, the APB morphed into the CAB (College of Applied Biology). In 2002, after years of hard work lobbying various MLAs, the organization achieved formal recognition from the BC government. With the advent of the CAB, biologists who applied to join had to provide proof of working for a required number of years in biology and a suitable professional report. When I joined the APB, they simply asked me to join, allowing me membership based on my reputation and a list of completed ecosystem mapping projects.

Later that summer, I engaged Terry Woods, one of our summer students from the Northeast Coal project in 1976, to help me with an ecosystem sampling project on the mid-coast of BC. The area we were to assess was TFL 41, which extended from just south of Terrace (the northern extent) to butt up against the Kitlope Heritage Conservancy in the south. It was managed by Skeena

Sawmills, now West Fraser Forest Products. TFL 41 eventually became part of the Great Bear Rainforest. In 2004, environmental groups ratcheted up their drive to protect a huge area of the BC coast. Some small portions were set aside as conservation areas. In 2008 Greenpeace, the Sierra Club, and ForestEthics came together, calling themselves the Rainforest Solutions Project, to apply added pressure on the BC government to pass legislation to make the protected area larger. Ultimately, in 2016, the Great Bear Rainforest was officially recognized. It encompassed a total area of 6.4 million hectares extending from the Discovery Islands in the south to the BC-Alaska border in the north. It is the largest, coastal, temperate rainforest in the world, and a large amount of it is now protected from logging. The logging that does go on is required to utilize an ecosystem-based management system.

Terry and I flew to Terrace where we met our helicopter that transported us to a guide-outfitters camp outside Kemano, on the west coast. At the camp, our hosts were Cy Ford and his wife. Cy was a crusty, grizzled, old hunter who seemed to enjoy killing bears. Terry and I had a separate cabin. Cy's wife prepared meals for us, which we ate with them in the dining room of their house. Upon entering the house, we immediately noticed small black, shrivelled bags about 10 to 12 centimetres long hanging from the ceiling. Upon asking what they were of Cy, he told us they were black bear gall bladders. Apparently, they were highly esteemed by Chinese apothecarists for their healing powers[5] (see Footnotes and Appendix for more information on bear bile).

One evening Cy decided to show us some video footage of bear hunts. The first scene we saw showed a pair of hunters in a boat that had been grounded against the shore. There was a large black bear above them on a steep slope. They shot the bear. It rolled down the hill to end up on the beach beside them. As if that wasn't bad enough, the next scene showed two hunters crouched behind a large log on an estuary. An unwitting grizzly was approaching them, and eventually, when the bear was close enough, the hunters jumped up with guns blazing at the doomed bear. Of course, the bear was killed. It was a heartbreaking event for me to witness. The life of a magnificent, highly intelligent animal snuffed out in an instant. The bear never had a chance, and the two hunters were so proud of themselves it made me feel sick. I could only bear (pardon the pun) to watch the bear 'snuff' film for about ten minutes, and I had to leave. Cy's wife took great delight in telling us about a forester who had been attacked and killed by a grizzly right in front of the helicopter and his companions.

Our field session passed uneventfully. Every day we were walking in and sampling spectacular, old-growth forests of western hemlock, Sitka spruce, and western redcedar. We saw incredible waterfalls streaming down rugged mountain slopes, emerald green and sapphire blue waters coursing towards the ocean, and sheer granite cliffs extending for thousands of metres. We saw a few bears from the chopper but never had any bear encounters on the ground. We also never saw any Kermode bears. The Great Bear Rainforest is home to the Kermode bear, sometimes called the spirit bear (*Ursus americanus kermodei*), a subspecies of the black bear[6] (see Footnotes and Appendix for more information).

Kemano, almost in the middle of TFL 41, is home to the Kemano Generating Station situated 75 kilometres southeast of Kitimat. Called Kemano I, it was built in 1954 to provide hydroelectricity for Alcan's Kitimat aluminum smelter. The powerhouse was built in a cavern created 427 metres inside the base of Mount Dubose. It was the largest producer of hydroelectric power in the province when it was made and is now the fifth-largest electrical plant in BC.

The Kemano I project was made possible by constructing the largest rockfill dam in the world at the time, the Kenney Dam on the east side of the Nechako Reservoir, 193 kilometres to the east. On the west side of the Nechako Reservoir, a 16-kilometre long water intake tunnel running through the Coast Mountains diverts the river water to penstocks for a huge 790-metre drop to the power station at Kemano. In 1995, the provincial government cancelled the Kemano Completion Project (Kemano II) that Alcan had planned since 1987.

Terry and I had an opportunity to take a tour of the generating plant. The sheer size of the facility, with its massive turbines and generators, was impressive. With my claustrophobia, being inside the bowels of a huge rock mountain made me more than a little nervous. I was relieved at finally leaving the underground plant to again breathe fresh air and see blue skies.

17 GRIZZLY BEAR HEAVEN

In 1989 I was part of a team of scientists tasked with collecting ecosystem data in the Khutzeymateen River drainage (about 44,000 hectares in total area) in northwestern BC. The Kateen River flows into the Khutzeymateen. 'Khutzeymateen' is a native word meaning 'a confined space of salmon and bears.' The Khutzeymateen River valley is located about 40 kilometres northeast of Prince Rupert on the coast. The valley had remained untouched except for some minor logging in the 1950s. In the 1970s, the talk of logging the pristine valley resurfaced. Our job was to gather ecosystem data that would be used to assess the habitat for grizzly and black bears. The valley was home to many bears of both species.

Field work was broken into three separate trips, and we were housed in a small cabin built on a barge. The barge was anchored at the mouth of the river within the confines of the estuary. From the cabin at the estuary, we were transported by helicopter or a small boat up the valley to run transects through the intimidating bush. The budget was limited, so our helicopter transport was restricted to being dropped off in the morning and then picked up at the end of the day as the chopper came in and out from Prince Rupert. The primary sampling team consisted of myself, my field assistant Terry Woods, MOF soil scientist Martin Geertsma, and MOF ecologist Sandra Thompson. The primary bear researcher was Grant McCutcheon. Jim Pojar and Allen Banner also joined us for one of our field sessions. Royal BC Museum botanist Dr. Bob

Ogilvie joined us for one session.

The valley has a wide floodplain (material deposited by the continual action of the water flowing down the river) with steep, rugged slopes and small creeks running into the main river. The bush was dominated by continuous thickets of devil's-club and salmonberry, both species with thorns, making hiking through them a significant challenge. The thickets were so dense at some locations we had no choice but to don raingear even in the heat of the mid-day sun to protect ourselves. Also present in the valley were stands of massive Sitka spruce, some greater than two metres in diameter and over 60 metres tall. These were the valuable trees that the loggers had been sizing up for removal.

Because of the large number of resident grizzlies, I always carried my shotgun. We were constantly making lots of noise as we moved through the forest, watching for bear signs. Bear signs came in a number of forms, including scat (we would assess it to see how fresh it might be), mark trails (which appeared as giant footsteps worn into the forest floor), mark trees (recognizable by the clumps of grizzly hair stuck to pitch oozing from small trees), and bear beds (large depressions, typically in sandy material on the floodplain, where bears would sleep when the weather was too hot). Terry, Sandra, Martin, and I spent a total of about 25 days collecting field data, and although we often heard bears moving about, we never had an actual 'close encounter.' I think we were lucky.

Grizzlies are the largest carnivore in North America. The coast of BC is home to many populations which have the advantage of salmon as a main food source. Feeding on salmon helps the bears to attain significant proportions. A large male grizzly bear can weigh in at more than 370 kilograms. The grizzly bear is named such because of the pale-tipped guard hairs that give them a 'grizzled' appearance. In the spring when the bears emerge from hibernation sleep, they need a fast, reliable food source. The river estuary offers them a smorgasbord, including sedges, horsetail, skunk cabbage, and tubers. They also harvest shellfish and crabs when the tide is out. Grizzly bears have a sophisticated sense of smell and acute hearing. Contrary to popular belief, they can also see very well, and their vision is comparable to that of humans. The black bears of the valley stay mainly in the side valleys since the grizzlies will kill and eat them if given a chance.

One day we were rushing to meet our helicopter pickup. We could hear the screaming of the turbine engine and the rotors cutting through the air as we traversed across the floodplain to meet our ride home. Terry and I were hurrying across a large, downed Sitka spruce, our caulk boots giving us ample traction. I made the mistake of grabbing a spur branch extending from the tree's main stem, sticking up in front of me. I used the branch as a lever and swung my body around it only to have the spur break, which threw me off the tree down from a height of about three metres to the ground below. I lay there with the wind knocked out of me, and luckily my backpack protected me from serious injury. It was a valuable lesson – never again would I rush to meet a waiting helicopter. The truth was that the pilot would never leave without us and would wait as long as it took for us to meet him. I was lucky that time not to hurt myself. I lay on the ground trying to catch my breath and eventually was able to right myself and continue to the chopper.

Bear spray was not readily available for field workers until the late 1980s, and research has shown that it is an effective deterrent against bears, even more so than a gun. Its active ingredient is Oleoresin capsicum which is derived from cayenne pepper. During our last field session, Allen Banner had an experience, which served us all with a warning regarding its use. Allen and his field assistant were waiting for a helicopter to arrive at the end of the day, and he decided to test his bear spray. He took it from its holster, released the safety and gave a couple of quick squirts. The capsicum spray shot out from the nozzle and formed a cloud in front of him. Without knowing it, he had gotten some of the spray on his hands, and when he went to relieve himself a short time later, he experienced an excruciating, burning sensation on his private parts. To reduce his suffering, he quickly grabbed a large wad of peat (sphagnum) moss as it was the only thing handy and stuffed it down his pants into the vicinity of his crotch. The sphagnum gave him immediate relief, and the occurrence became known thereafter as the 'sphagnum poultice' incident.

In 1994 the BC government established the Khutzeymateen Provincial Park (and Grizzly Bear Sanctuary) to protect the magnificent forests and grizzlies of the valley. It was a wise decision that prevented logging which would have forever altered the pristine environment. Today, there are two charter companies that are permitted to take limited adventurers into the valley to view and observe the grizzlies living in their habitat. The Khutzeymateen was one of my all-time favorite field projects, and I felt privileged to have had the opportunity to work there and experience its savage wildness.

SECTION 4

Shearwater Mapping Ltd.

Anne and I started the company to capitalize on the growing need for Geographic Information System products (computer-generated maps). We thought we would get rich, but that was a myth. As the company grew, we hired more staff to allow us to take on bigger projects. We specialized in Terrestrial Ecosystem Mapping, and most of our projects were in the vast wilderness areas of BC. We became experts at doing helicopter surveys and producing quality ecosystem maps for an array of clients.

19
THE BEGINNING OF SHEARWATER MAPPING LTD.

In 1990, Anne and I formed Shearwater Mapping Ltd. One of our primary focuses was to enter the GIS (Geographic Information System) era and provide digital maps to clients. We started out with about $10,000 worth of the necessary software and hardware to create sophisticated digital map products. John Bunning was an old friend who had graduated from BCIT with me in 1970 and was working for a company in Vancouver responsible for updating forest cover maps for the MOF. In 1989 I had volunteered Anne to help him in a job to update provincial forest cover maps. Anne had no formal training in GIS, but I thought she would enjoy the challenge due to her propensity for being extremely proficient with computers and computer programs. It turned out I was right – she was a whiz! She had the right kind of personality and attitude to excel in the demanding world of GIS where the creation of digital files required an exacting commitment to accuracy and the need for following detailed prescribed standards. Anne served as the president of the company, looking after the finances, GIS work, and human relations. At the same time, I ran the field crews, being responsible for pre-typing aerial photos, fieldwork, mapping on photos, and report writing.

We started out working on the MOF forest cover map files with three employees, Lynn Brunsdon, Grant Hamilton, and Glen Beckwith. Eventually, we graduated to completing other digital map projects for other government agencies and private companies. During 1990, as we were gearing up on the forest cover mapping program, the MOE published a document entitled Describng Ecosystems in the Field. It was a field guide that provided a 'bible' for field workers on how to collect and record field data for environmental conditions, surficial material, soil, vegetation, mensuration (forestry measurements), and wildlife. The guide was updated in 1998 (Field Manual for Describing Terrestrial Ecosystems) and again in 2010 (Field Manual for Describing Terrestrial Ecosystems, 2nd Edition). In 1990, at the same time as when the original field guide was produced, the province adopted a standardized methodology for producing ecosystem maps called TEM (Terrestrial Ecosystem Mapping), which included a manual for mapping and the creation of standardized field data collection forms. At the time of their introduction, these materials resulted in a progressive, clearly articulated, accepted way of collecting environmental field data and producing ecosystem maps. The data and maps could be used for a variety of purposes, including the assessment of forest productivity and wildlife habitat and any proposed activities that could potentially impact the natural environment.

In the summer of that year, a young biologist by the name of Brent Keeping from Newfoundland wandered into our office one day looking for a job. His accent was so strong that I had trouble understanding him, but he was keen and willing to do anything to learn about the environmental business. Anne and I liked his countenance, energy, and pluck, so we hired him to be my field assistant and help with the digital mapping. Brent turned out to be the best field assistant I ever had, staying with Shearwater until 1998. Little did we know it in 1990, but Shearwater Mapping Ltd. would grow to be a company with up to 18 employees specializing in TEM projects for the next 12 years.

THE SNIP GOLD MINE CAMP

In 1991 Shearwater took a contract with the MOE, Wildlife Branch in Smithers to complete a biophysical survey in the Iskut River and Lower Stikine River drainages in northwestern BC. Ron Kowall, a government soil scientist I had known since 1970 during the Kelowna days and Brent accompanied me into the wilderness to collect ecosystem data. We caught our helicopter in Smithers and flew to a gold mining camp located in the lower Iskut River valley. Our pilot was a large, jocular fellow by the name of Tom. The camp was at the Snip Gold Mine owned and operated by Cominco Ltd. Between 1991 and 1999, when it closed, the mine produced about one million ounces of gold.

Tom and the chopper stayed with us at the camp, and he flew us out to our sample sites every day. He was there to provide us with an expensive taxi that made our sampling efficient and relatively easy. In those days, the JetRanger chargeout rate was around $750 per hour plus fuel costs. Today the same machine goes for about $1,200 per hour. Choppers are not cheap, but they are the most effective way to get around and access sample sites for wilderness environmental work.

Until this point, I had not had many opportunities to stay in camps, either for logging or mining operations. The Cominco camp on the Iskut River was amazing and probably the most luxurious one I have ever stayed in. We all had our own private room with a large bed and individual TV. The dining room was spacious and offered a huge variety of different types of food, and the staff was friendly and courteous. Brent remembers that the room doors had gold numbers, and in the dining room, the desserts were kept in tall fridges with glass doors. Interestingly, even today, camps are essentially dehumanizing places with workers only there to sleep and eat between workdays. They may be newer and modern, but the workers are, for the most part, institutionally housed with strict rules and security guards everywhere.

This project was the first venture into the mountains and wilderness of BC for Brent, and he had his eyes opened wide one day to the potential dangers and risks. We had disembarked from the chopper at one location, and upon entering the forest nearby, we came across a large subalpine fir with grizzly claw marks on it. The claw marks were easily three metres off the ground, and I can still see the look on Brent's face as we regarded the claw marks left I took a picture of Brent beneath the tree with him holding the shotgun, and he looked scared and shocked. I am sure the experience had a lifelong impact on him.

Cominco had a unique way of transporting their gold ore from the mine site. Large sacks of ore were loaded onto a huge hovercraft, able to operate for eight months a year, which then travelled down the Iskut River and the Stikine River to tidewater where the ore was transferred onto a large ship to eventually arrive in Japan via Wrangell, Alaska.

One day, when we were back in camp a bit early, we were given a tour of the underground mining operation, and it was illuminating in the revelation of the process of cleaving the gold ore from the inside of the mountain where the deposit was located. I still remember being in the darkness and driving around in a large-wheeled transport vehicle. With my fear of being in confined spaces, the experience made me quite agitated. When we finally emerged from the dark pit, I was indeed relieved.

The trip passed without any bear encounters, although there was ample sign of their presence every day, and we returned home to Victoria none the worse for wear.

20 GUIDE OUTFITTERS, THE MISTY ISLES, AND WORKING IN THE SNOW

1992 was a busy year for Shearwater. The first project we worked on was the Northern Rockies Wildlife Habitat Mapping project for the MOF. They were interested in determining the effects of prescribed burning. These were areas typically burnt in the spring or fall to help maintain the production of grassland ecosystems on warm slopes of the northern Rockies. Denny, Brent, and I spent two weeks sampling the varied ecosystems of the Rockies. During fieldwork we stayed in two guide outfitters facilities: one near Toad River and the other in the Muskwa River valley.

We again had the luxury of having the chopper and pilot with us for all field sampling for this project. We surveyed a vast area in northeastern BC with spectacular landscapes, including the low elevation, dry grasslands, as well as boreal forests and alpine areas in the Rockies. As we flew over the tracts of wilderness, we saw extensive herds of Rocky Mountain elk and groups of thinhorn sheep and mountain goats. Thinhorn sheep are different from their cousins, the Rocky Mountain bighorn sheep, in that the rams have thin horns with far less curl than the bighorns. Thinhorn sheep spend much of the year roaming the high cliffs and crags of the mountains.

During our time at the guide outfitters, we were able to sample every imaginable form of wild game, including moose, venison (deer), sheep, goat, caribou, and elk. At the Toad River location, run by Dave and Elly, we were treated royally. Brent still remembers being served dinner by Elly's adult daughters as he was infatuated with them. The other camp, located on the Muskwa River, was run by Gary Vince and his family.

The guide outfitters of the region served to provide hunting services for dedicated hunters who came from all over the world to try for trophy animals of different species. In 2016, the guide outfitting industry in BC was responsible for generating more than $92 million of economic impact from more than 10,000 hunting clients[7] (see Footnotes). BC can boast of hosting 18 of the 30 'big game' species found in North America. In 2017 the government of BC imposed a ban on the hunting of grizzly bears in the province. Research has shown that viewing and eco-tourism opportunities associated with grizzlies are far more productive economically than hunting these wilderness icons.

Later in the summer, I would participate in one of my all-time favourite projects. Pedology Consultants had secured a contract to conduct a biophysical survey of South Moresby Island, the most southern island of the Queen Charlotte Islands. The islands, consisting of the two main islands of Graham Island and Moresby Island, have been the traditional home of the Haida Nation for centuries. Pedology Consultants had been engaged to complete biophysical mapping of south Moresby. Denny Maynard and I were hired as subcontractors to work on their behalf as the project's terrain mapper and vegetation ecologist.

Following deglaciation, about 12,000 years ago, the islands were settled by the Haida. Estimates of their population at the time of colonial contact in 1774 were around 30,000 people. With the onset of resource extraction in the 1800s, especially the taking of sea otter pelts, the population was decimated by smallpox and other introduced diseases. By the early 1900s, the Haida had shrunk to only about 600 people. Today about 2,000 native Haida are living on the islands.

During World War II, Sitka spruce was heavily logged on the islands, being used primarily for plane construction due to the lightness of the wood. I remember talking to the owner of the Queen Charlotte Hotel and listening to him tell me stories of his dad cutting down trees that were over three metres in diameter. All the spruce that size are long gone now. Since the 1940s, most of Graham Island, the northern island, had been heavily logged with very few stands of natural, large Sitka spruce remaining. In 1974, disputes came to bear on the future of Moresby Island. The conflict continued until 1993, when the name Gwaii Haanas replaced the colonial moniker[8] (see Footnotes and Appendix for a detailed chronology). Haida Gwaii is also known as the 'Misty Isles.'

Our field team of Denny, Bobby Hammersly, Richard Thompson, and I flew from Victoria to Sandspit, then caught the ferry to Queen Charlotte City, where we met Keith Rowsell and his wife, Barb. Keith was the builder and captain of the Anvil Cove; an aluminum-hulled schooner used to access the waters around Haida Gwai. It would be our means of circumnavigating the island to access sample sites to collect vegetation, terrain, and soil data. The Rowsells still run their charter boat business out of Queen Charlotte City. Keith had the weather-beaten, lined face of a seasoned mariner; Barb looked like she could be the mother of everyone. After exchanging introductions, we stowed our gear aboard the ship and cast off the lines, then headed west through the narrow channel between Graham Island and Gwaii Haanas. It was a beautiful sunny day with a slight breeze, and it was pleasant to sit on deck watching the shoreline pass by with its endless changing pattern of vegetation, coastal bluffs, and tidal environment.

We would be shipmates for two weeks, completing a counterclockwise route around Gwaii Haanas. There was room for the four of us to sleep in the main part of the ship, but it took a while to adjust to the cramped kitchen and galley. The head (toilet and sink) was right beside the galley, and it was challenging to use the head while people were literally sitting only a few feet away in the galley.

Within a few hours, we had reached the open ocean of the west coast. Here the Pacific Ocean pounds the shoreline mercilessly. The average annual precipitation along this coast amounts to 5,000 millimetres, with a portion of it coming in the form of a near-perpetual fog. The environment is generally called the 'fog zone'. Steep slopes that run right into the salt chuck are populated by stunted shore pine (a cousin of lodgepole pine that we called 'bonsai trees' or 'popcorn trees') and a variety of moisture-loving plants including sedges, sphagnum moss, and sundew (an insectivorous plant). At regular intervals we would see strips of vegetation and soil torn from the slopes and folded like accordions at the interface with the beach. The ground becomes so saturated with the endless precipitation that gravity tears the surface layers away like giant Slurpees.

The sea was rough, and the constant pitching of the ship soon caused Bobby and I to feel the effects of sea sickness. I managed to overcome my biliousness, but poor Bobby spent much of the first day in her bunk feeling nauseous. That night, and every following night, Keith found a sheltered spot for us to anchor. Barb prepared a mouth-watering feast of fish, rice, and vegetables. When we finally hit the sack, we drifted off to sleep with the sound of the gentle waves lapping at the hull of the ship.

In the morning we had a hearty breakfast of eggs, ham, and fried potatoes, then loaded our gear into the Zodiac. Keith ran us into shore, where we disembarked and headed off to collect our field data. At the end of the day, Keith or Barb picked us up and took us back to the Anvil Cove. We spent just a few days on the outer coast because sheltered bays were limited and eventually arrived at the southern tip of Gwaii Haanas where Anthony Island is situated.

Anthony Island is home to the abandoned Haida village of

Ninstints which is now a UNESCO World Heritage Site. We spent time there marvelling at the ancient, weathered poles. Many of the poles were in an advanced state of decay with some being propped up with supports.

At the very most southern tip of Gwai Haanas is Cape St. James where an abandoned lighthouse stood as a reminder of a bygone marine safety network. The lighthouse became automated in 1992 (September), and for many years the Department of Fisheries and Oceans considered de-manning all the lighthouses on the east and west coasts of Canada. In 2011 that plan was rejected. As of 2018, only 27 of the 54 lighthouses in BC were manned. We continued away from the outer coast and into smoother waters. The east side of Gwaii Haanas is vastly different from the west coast with much less precipitation and a markedly different set of ecosystems.

We had a few days where Keith would drop us off on the shore and we would meet a helicopter for conducting high elevation fieldwork. It wasn't uncommon for us to see black bears foraging in the intertidal zone for food, but there are no grizzlies, cougars, or wolves on the islands. The black bears of Haida Gwaii are the largest not only in BC but in North America. They are a distinct subspecies with larger jaw and cranium than the typical species. The intertidal zone offers them a rich source of food, including sea urchins, crabs, seaweed, and carrion. The larger male bears of the islands can weigh more than 300 kilograms.

On our excursions inland, we would stop to collect soil and vegetation data at selected sites. Denny and I nicknamed Richard as 'Lumpy' due to his general large countenance and shape. He and Bobby were along as field assistants and one of their main tasks was to dig a soil pit at each sample location. At one location, Richard was having trouble digging through the layer of surface tree roots that always seemed to be present. After a bit of feeble digging, he started to whine and complain at how hard it was. I lost my cool and blew up at him, grabbing the shovel from his hands. In a furious display of hacking and bashing with the shovel, I was able to break through the roots and reach the mineral soil beneath. He never complained again. Richard was not particularly agile and would fall every day at least a few times, tripping over protruding roots or branches laying on the ground.

On the east coast of the islands, the vegetation was severely modified by the presence of deer. Ten species of animals have been introduced to Haida Gwaii, including deer, rats, and raccoons. Sitka black-tailed deer were introduced five different times to the islands by the BC Game Commission between 1880 and 1925. Since that time, in the absence of competition and significant predation, the deer have proliferated. The only natural predators are the bears, and they take few deer. The effects have been drastic, with many forested sites lacking shrubs and herbs due to intense browsing by deer. Residents are allowed to take up to 15 deer per year, but even the generous hunting quota has done little to reduce the effects of deer. I even observed deer browsing on young Sitka spruce (a sharp-needled tree species), which I have never seen anywhere else.

We continued up the east coast and came to another abandoned village at Tanu. The long weather-beaten houses were covered with moss. It was sad and eerie to walk through the site, knowing that most likely everyone who had lived there had been a victim of smallpox. The place was heavy with ancient memories, desolation, and bad spirits. We were happy to leave and continue north.

Towards the end of our trip, we spent three days anchored near Hotspring Island. In the evenings, Keith would run us ashore, and we would walk to the amazing hotsprings that consisted of two pools very close to the crashing waves. The water was therapeutic, and we sipped beer and relaxed in the soothing mineral waters as the sun set. Many kayakers travel to Gwaii Haanas annually to travel the east coast, and Hotspring Island is a favorite stopping point.

It had been an amazing trip and a privilege to experience Gwaii Haanas from a marine and land perspective. We had seen whales, sea lions, porpoises, bears, as well as numerous bald eagles and ravens. The raven has been my favorite bird for a long time. They are part of the large group of Corvids (*Corvidae*) that include ravens, crows, jays, and magpies which are among the most intelligent of birds[9] (see Footnotes). The common raven that is found all over Canada and most of the western United States can live for more than 20 years in the wild. They are masters of the sky, and I would often see them flying as mated pairs performing remarkable aerobatics. Their wingspan can be large (as much as 150 centimetres). Ravens have between 15 and 30 different types of vocalizations and can even mimic the human voice. They mate for life and have achieved better than dogs and cats in intelligence tests. They have the largest brain-to-body ratio of any bird, as illustrated by their feeding practices, memorization abilities, the use of tools, and advanced social interaction. Ravens also have an esteemed place in Haida mythology and culture[10] (see Footnotes and Appendix for more information on the raven in mythology). Totally white (not true albinos) ravens have been seen on Haida Gwaii. Colin (my son) and I were fortunate to see two such mythical birds near Port Clements in 1996.

We were sad to arrive back at Queen Charlotte City and bid goodbye to Keith and Barb. Little did I know that over the next ten years I would be lucky enough to revisit the Misty Isles many times.

In September, I had a contract with the BC Ministry of Environment to complete a biophysical inventory program in the McGregor River and Herrick Creek drainages on the west slope of the Rockies. My partners were Larry Lacelle, a government soil scientist who passed away in 2013, and Mike, a contract wildlife biologist. It was late in the year for vegetation surveys, and we knew that the weather might be a concern in the mountains.

The McGregor River originates in the Kakwa Provincial Park and Protected Area, generally flowing west and northwest to join the Fraser River northeast of Prince George. Larry and I drove to an empty logging camp to hook up with Mike and our helicopter pilot. The study area was home to significant populations of moose, woodland caribou, grizzly and black bear, deer, and grey wolf. The western flank of the Rocky Mountains formed the eastern boundary of our area.

Other than the three of us and the pilot, the camp was deserted except for a cook who prepared our meals. The weather started out with cool days and endless rain; it seemed as though I was wearing rain gear and rubber boots every day. A constant layer of thick, oppressive clouds made flying difficult. On top of the deteriorating weather, we were met with extensive areas of devil's-club, which is related to ginseng. Devil's-club, which can grow up to an extraordinary height (I have seen it as tall as six metres in the West Kootenays), is aptly named. The stems, up to three centimetres in diameter, which are covered with coarse, woody spines, are irregular in form, allowing them to be flexible. Sometimes when walking through them, the stems would lash us on our hands. It sometimes takes months to extricate the spines from the skin. Even the large palmate leaves (up to 40 centimetres across) have spines along the leaf midrib and veins. The plant has been used by First Nation groups (Tlingit and Haida) as a traditional medicine for diabetes and rheumatoid arthritis. Bears are known to feed on the bright, red clusters of berries that develop at the end of the stems.

After the first week of work, it began to snow, making our sampling in the mountains even more difficult. As we flew along the warm, west-facing slope of the Rockies, we began to see piles of recently disturbed earth. Not knowing what had caused them, we landed at one such site to investigate. Lo and behold, the piles of soil were located in meadows that contained large numbers of avalanche lily. The grizzly bears were targeting the plant's starch-rich tubers to add to their fall bulk-up menu before heading into hibernation. By following these dig sites in the chopper, we came across numerous foraging grizzlies. In one day alone, we noted five magnificent animals feeding exclusively on the lilies.

It was a challenge to do vegetation plots in the mountains with more than 20 cm of snow obscuring the plants. Somehow, I found a way to collect relevant plant data by scraping snow away and identifying the dead and dying plants. I was operating in the realm of forensic plant ecology. Many times during my career, I was called upon to be able to accurately determine plant species in the late fall. It was a useful skill to have.

21

THE GLACIER VIEW MOTEL

Denny, Brent, and I would resume our field relationship on the Tweedsmuir-Atnarko Grizzly Habitat mapping project in 1993. We made the long drive from Victoria to Williams Lake and then made a turn west on to Highway 20, driving another 437 kilometres to Hagensborg in the Bella Coola valley. For two weeks we drove different roads in the study area and used some helicopter access to sample the ecosystems of a portion of Tweedsmuir Provincial Park and the Atnarko River drainage. The study area was home to many grizzly and black bears, and my trusty companion, the shotgun, was at my side whenever we ventured into the bush.

I don't have any vivid memories of field experiences on this project except for encountering some massive Douglas-fir trees that were more than two metres in diameter.

The most memorable moments from this trip all relate to the Glacier View Motel, which consisted of a set of decrepit, rundown buildings that had seen better days. There was a large 'For Sale' sign in front of the main office. Rumor had it that the sign had been there for more than ten years. The 'glacier' was a small piece of ice that hung in a mountain gully high above and to the east of the motel. Our room was in an old clapboard building, with bare wooden floors, rusty bed frames, and aged mattresses. Brent remembers that the thin, stained mattresses lay right on top of the springs. The room was never cleaned during our two-week stay, and if we needed toilet paper, we had to ask for it at the front desk.

The place was run by an old couple whom we nicknamed Ma and Pa Kettle, after the characters in a series of comedy films released in the 1940s and 50s by Universal Studios. Just like their movie counterparts, they were always arguing and disagreeing on how things should be run and what needed to be done.

The first night we were at the motel we went to the dusty, old restaurant and perused the cracked, faded menu. When Ma came to take our order and we told her of our meal wishes, she brusquely told us that those items weren't available and fresh salmon was our only choice for an entrée. So, we had the salmon, which sure enough was fresh and tasty. The only problem was that salmon was the only item available for dinner for the rest of our stay. We noticed during our time there that the salmon steaks got smaller and smaller, which meant they were probably being cut from the same fish. Pa wore a pair of jeans that always seemed to be just barely staying up. As a result, when he bent down to do any work, his ample 'crack' was visible for the whole world to see. Not a pleasant image one wanted to see while waiting for a meal. When we finally left the Glacier View Motel, we were happy to be on the road back to civilization. Over my career I have stayed in some wretched camps and motels, but the Glacier View had to be ranked close to the top of the list for poor and uncomfortable accommodations.

22

SWIMMING IN THE FRASER RIVER AND THE TALUS SLOPE

1994 found me working on the Dog Creek Biophysical Project, collecting ecosystem data in an area west of Williams Lake, including a section of the Fraser River and extensive natural grasslands. Dog Creek is an unincorporated ranching settlement located about 100 km south of Williams Lake. Located on the northeast side of the Fraser's confluence with Dog Creek, it is opposite the historic Gang Ranch on the west side of the river, connected by an old suspension bridge serving local ranch roads. Dog Creek's first post office opened in 1873. The ranches at Dog Creek and the Gang Ranch hired mostly First Nations cowboys, most of them coming from the Canoe Creek/Dog Creek Band. The Gang Ranch is one of the oldest and biggest functioning ranches in western Canada, with a total area of over 12,000 hectares. Surrounded by a rough and unforgiving landscape, the ranch is a mosaic of lush hay fields and native grasslands. The main ranch is a picturesque display of heritage buildings surrounded by hitching posts and massive black cottonwood trees.

For the fieldwork, I was joined by Polly Noakes (now Polly Uunila), a terrain mapper with June Ryder and Associates, and Debbie, a grassland expert whose last name escapes me. We were supposed to be joined by Gina Roberts, a wildlife biologist, but at the last minute, before coming into the field, she developed a case of the hantavirus (formally know as hantavirus pulmonary syndrome) which almost killed her. The disease is carried by rodents that shed the virus in their urine, droppings, and saliva. The virus is mainly transmitted to people when they breathe contaminated air. The deer mouse, which is common throughout most of North America, was probably the culprit in her case.

We stayed at a set of cabins not far from Dog Creek and did all our work on foot after driving to selected accessible locations by road. At the beginning of the project, we were joined by Trevor Goward, an expert lichenologist. Like myself, Trevor had started out in one career with a degree in classic English literature and ended up in another. He was regarded as one of the pre-eminent lichen experts in BC, if not Canada, with numerous new species discovered and scientific publications about lichens to his credit. Trevor spent a few days with us tramping about and educating us on the small lichen species to be found deep down between the clumps of grass. Many of the species were members of the Cladonia and Peltigera genera. I remember in particular one 'cute' little lichen by the name of *Ochrolechia upsaliensis*. This tiny organism looked like small white donuts sitting on litter and soil beneath native grass species.

Later in the project Mike Sarrell, a herpetologist (reptile biologist), joined us for a few days to assess reptile habitat and to look specifically for rubber boa specimens. Rubber boas are found from southern BC to as far north as Williams Lake, occurring in grasslands, coniferous forests, dry pine forests, juniper scrub, and riparian areas. The species name suits this curious creature well. Between 35 and 80 cm long and thick-bodied, the boa resembles a rubbery tube with a slight taper at each end – similar in shape to what you'd roll out from a ball of clay[11] (see Footnotes and Appendix for a detailed account of the rubber boa). I have no recollection as to whether Mike did find any boas, but he did locate numerous lizards and skinks. Rattlesnakes are common in the dry interior of BC, but they don't extend into the Dog Creek area.

During fieldwork, the weather was extremely hot and oppressive. On a couple of occasions when we were near the Fraser River, we took the opportunity to strip down to our underwear and go for a dip in the mighty river. The Fraser River is the longest river within British Columbia, rising at Fraser Pass near Blackrock Mountain in the Rocky Mountains and flowing for 1,375 kilometres into the Strait of Georgia at the city of Vancouver. It is the 11th longest river in Canada and discharges almost 20 million tonnes of sediment into the ocean annually. It was named after Simon Fraser, an explorer of the 1800s. From an economic perspective, it is the most important river in BC, having formed the Fraser delta that comprises almost the entire Fraser Valley with some of the most fertile, rich agricultural lands in North America. The river is host to the white sturgeon, which can grow up to four metres in length, weighing more than 500 kilograms, and five species of Pacific salmon: chinook, chum, coho, pink, and sockeye. Our dips into her dark, muddy waters gave us a firsthand experience of her sediment loads. Upon leaving the rushing water, our underwear was full of fine sand and silt.

Also, in 1994 we were asked by Bill Beese, forest ecologist with MacMillan Bloedel Limited, to participate in a small TEM project as a pilot for mapping of their TFL (Tree Farm License) lands on the coast of BC. After completing the work, we were selected, starting in 1995, to be the company responsible for TEM work on all their TFL and private forestry lands. Their TFL holdings included TFL 39 (near Powell River) and TFL 44 (Haida Gwaii). MacMillan Bloedel Limited, sometimes referred to as 'MacBlo', was a Canadian forestry company headquartered in Vancouver, British Columbia. It was formed through the merger of three smaller forestry companies in 1951 and 1959. Those were the Powell River Company, the Bloedel Stewart Welch Company, and the H.R. MacMillan Company.

The company was created in 1919 by Harvey, or H.R. MacMillan, British Columbia's first Chief Forester. His time as Chief Forester gave MacMillan the opportunity to scope out the vast forestry resources of coastal BC. MacMillan reportedly gained considerable experience in world lumbering during World War I. With his colleague, Whitford Julian VanDusen, another forester, MacMillan incorporated a company in 1919 to sell British Columbia lumber products to foreign markets.

In 1924, they established a shipping company that would become one of the world's biggest charter companies. With the creation of Seaboard Lumber by the other mill owners in British Columbia, there was a major threat to MacMillan, as Seaboard was to export all the lumber from the companies that founded it leaving MacMillan without the lumber needed to fulfill their orders. MacMillan responded by beginning to purchase mills and creating the first truly integrated forestry company in British Columbia. MacBlo was bought by Weyerhaeuser of Federal Way, Washington in 1999. TFLs, which came into being in 1956, guaranteed logging rights to selected forestry companies for renewable (based on successful performance and the provincial government's acceptance of forest management plans) terms of 21 years.

Colin and I did a small mapping job near Pemberton, BC at the end of the summer. I can't remember the name of the company or the actual area that we were working in, but the job was significant because we had an accident that almost killed both of us. One day, we parked on the side of a logging road and needed to access a stand of Douglas-fir above us. Between the stand of trees and us was an area of talus running upslope for about 100 metres. Colin and I carefully began picking our way through the jumble of large boulders covered in lichens and mosses. Eventually we arrived near the top of the talus when suddenly the boulders beneath us started to move. In an instant, the entire slope gave way beneath us and only by pushing and throwing Colin to the side and scrambling into the trees myself did we survive. A huge cloud of choking dust enveloped us, and we gasped for breath as we realized how close we had come to certain death. I had always thought that talus slopes were stable, but that day my opinion changed.

THE MACHMELL CAMP AND THE MAGICAL ALDER TREE

In 1995, Shearwater embarked on a seven-year TEM program for MacBlo. Our first project was in the Schoen-Strathcona drainage on northern Vancouver Island. That year we completed ecosystem inventory and mapping for the Schoen-Strathcona (northern Vancouver Island), Machmell-Sheemahant (mainland of BC, mid-coast), Louise Island (an island east of the main part of Haida Gwaii), and Flores Island (west coast of Vancouver Island).

In 1991 when the NDP (New Democratic Party) came to power in BC, they initiated the FRBC (Forest Renewal BC) program, which allocated funds to forest companies to carry out a number of forest management activities to improve the way logging was done in BC. Our TEM work qualified as one of these activities; hence companies like MacBlo were able to access money to complete ecosystem mapping programs. Our work was used by the company as input to their forest management plans to identify forest ecosystem networks and critical ungulate winter habitats. That summer, we also did work for Pacific Forest Products in the Machmell River and Sheemahant River drainages.

Our crew for these projects included Brent and me, as well as Ron Kowall, Rod Dalziel, and Tyler Innes. Ron was an excellent soil scientist and terrain mapper who had worked for the provincial government in Kelowna when I worked there in the 1970s. Still employed by the government, he had moved to Victoria around 1976 but had been declared 'redundant' in 1994. Rather than toil away in the meaningless job they had transferred him into, he decided to retire from the civil service and come to work with Shearwater. Tyler and Rod were recent graduates of Camosun College in Victoria. Tyler worked with us until 1996 when he left to continue his career with other companies. Tyler now works for Jacobs, a large multinational engineering and environmental company. Brent now lives in St. John, Newfoundland, working as an Impact Assessment Project Manager with the federal government. Rod is a senior park ranger on the Sunshine Coast. Ron stayed with Shearwater until when we closed the company in 2002. For our work on Haida Gwaii, we were joined by Mitch Williams, a young Haida worker who was full of mischief, and Colin.

For our work with MacBlo on Vancouver Island and the adjacent mainland coast, we used FIFT (Forest Industry Flying Tankers) as our helicopter source. FIFT was a company formed by several major forest companies operating on the coast of BC, including MacBlo, TimberWest Forest Corp., and Pacific Forest Products. FIFT came to be in the 1950s due to a significant number of large forest fires. The forest companies decided they could fight fires more efficiently and economically if they controlled their own water bomber fleet. They acquired four Martin MRM Mars flying boats which had originally been conceived and built by the US Navy starting in 1938. The flying boats were massive machines, over 36 metres long with a wingspan of 61 metres. By the time FIFT had converted them to water bombers, they could carry and drop on a fire over 27,000 litres of water. Along with the water bombers, FIFT also owned a number of LongRangers, which were bigger and more powerful than the JetRanger, as well as a Grumman Goose. The LongRanger (206L) had an extra row of backward-facing seats in the passenger compartment allowing for the transport of up to six passengers, as well as one passenger up front and the pilot. The Grumman Goose was an amphibious transport plane that first flew with the US Navy in 1937. We used the 'Goose' occasionally to transport our crews into logging camps. The helicopter pilots were all skilled and highly accomplished, providing an essential level of comfort and safety. Using FIFT meant we didn't have to arrange for charters with the normal commercial companies like Okanagan, Canadian, or Vancouver Island Helicopters. In 2001, FIFT became a wholly-owned company of TimberWest Forest Corp.

I have, on occasion, flown with a helicopter pilot who unnerved me. One such incident happened in the Williston Lake area when I was flying with Larry Lacelle and Lynn Giguere, a biologist with TimberWest. Chuck, the pilot, made us all nervous with his flying, and the situation became unbearable one day when we were preparing to land on an alpine ridge. Larry was sitting beside the pilot, and as we were about to set down, Chuck turned to Larry and said, with his eyes protruding, "Which way is the wind blowing"? It is always the job of the pilot to know what direction the wind is coming from. That was it. At the end of the day, we contacted the helicopter company and told them we needed a new pilot. Thankfully, our request was granted. I always figured if the pilot was old and had grey hair, then we would be fine. The seasoned pilots tended to have more experience flying in the mountains and were generally safer. Besides that, they had no need to impress you with their daredevil skills.

When Rod, Tyler, Brent, and I arrived, by chopper, at the Machmell logging camp, we were told we could not stay in the main bunkhouses but were to be staying in a rough, old cabin nearby. We were fine with that since we didn't have to deal with the loggers and had our own bedrooms and bathroom. Little did we know that the cabin was infested with many deer mice. On our first night in the cabin, we were awakened by the small rodents scurrying about in the dark and even running over our beds. Being concerned with the risk of the hantavirus we set about to resolve the mouse problem. Brent had some experience in the past dealing with the creatures and devised an ingenious way to trap them. He took a large tub, filled it to near the top with water and then suspended a string, pulled taut, across the top of the tub from one side to the other. In the middle of the string was an empty toilet paper roll which he smeared with peanut butter. The mice would be attracted to the smell of the peanut butter, run out on the string where they fell off the free-wheeling roll into the water where they drowned. It wasn't humane, but it did the job. In the morning after the first night of the trap's implementation we found three dead mice floating lifelessly in the tub. Every night we were there we would catch a few mice and gradually their nocturnal intrusions declined.

That summer there was an extreme fire hazard on the coast, and we had to get out of bed at 4 am to eat breakfast and make our lunch. Since it was still dark at that time of day, we had no choice but to go back to bed and then get up again around 6 am to go to work. We had been warned by the camp staff that three young grizzlies were frequenting camp. It wasn't unusual for us to see them in the morning as we walked from our cabin to the dining room.

Our pilot, Dennis, was an unfriendly, surly fellow. He took great delight in 'dusting' us off when he landed on logging roads to pick us up. One day at camp, he pretended to be a bear by growling outside the cabin and scratching at the door. I almost shot him, grabbing the shotgun and preparing to kill the 'bear' trying to get us. He thought it was funny, but he never pretended again to be a bear.

It was Rod's first foray into the territory of grizzlies, and I remember the day I showed him the obvious bear signs, including mark trails and mark trees. His eyes grew wide as I explained to him how and why the bears created these obvious indicators of their presence and territory.

On Louise Island, we stayed at the MacBlo logging camp and were given the use of a 'crummy' to travel around the island to various sample sites. A crummy is a vehicle (a small van, bus, or railway car) used to transport loggers or other resource workers from camp to the various active logging areas. Our crummy had seen better days, but we had no choice other than to use it. The wretched thing was in the order of eight metres long, and on numerous occasions, we got it wedged in the middle of large ditches that had been dug across logging roads to channel water to the roadside. On one occasion, it took the five of us hours to dig the crummy out from one such ditch.

Mitch had a bit of the 'raven' in him. We were leaving a sample site one day and came back to the crummy to find him sprawled on the steps of the vehicle, thrashing about and foaming at the mouth. He pretended that he had eaten Pacific yew berries, which are poisonous, with a few of the bright red berries lying scattered about him. After a few seconds when we tried to revive him, he came to life, sputtering and spewing berries out of his mouth. As he laughed heartily, he told of us the 'joke' he had played on us. We chastised him and told him it wasn't at all funny, and we didn't appreciate his humor.

On Gwaii Haanas, there were Haida Watchmen stationed at five ancient village sites, including one on Louise Island[12] (see Footnotes and Appendix for detail on the Watchmen). On the east side of Louise Island, where Skedans is located, Mitch's aunt and uncle were working as Watchmen in 1995.

One day when we had some extra time, we parked the crummy and walked for about a km down a winding, steep trail to the village site below. Mitch's uncle spent time telling us Haida stories and talked about the poles still standing nearby. At the end of his talk, he told us to follow a trail leading out of the village to see something wondrous that we would truly appreciate. He wouldn't tell us what it was, but he said it was 'magical.' With our curiosity piqued, we took his advice and walked to the edge of the site and there before us was something truly magical and astounding. It was a gigantic red alder tree of massive proportions.

Red alder is a common tree found only in coastal parts of

BC that acts as a 'pioneer' species to repopulate disturbed sites, including logged areas. It grows rapidly, maturing in 40 to 60 years, and usually the largest specimens can grow to a maximum diameter of 75 centimetres with a height of 30 metres. The trees fix nitrogen in the soil, thus improving site fertility for evergreen species that establish in their shade (alder does not grow well in the shade). This tree was over two metres in diameter, with numerous secondary branches rising like tendrils from the main stem. On the secondary branches were many epiphytic mosses and licorice fern. We were amazed and stood speechless taking in the enormity and immensity of the beautiful giant. Using a diameter tape, which we usually used to measure tree size on our sample plots, we measured the tree at 223 centimetres.

After taking many pictures, we left the site, revisited with Mitch's aunt and uncle, and then headed up the trail to the crummy. I was lucky to go back to the site with Anne in 1997 to show her the special tree, and she was duly impressed. Bill Beese at first couldn't believe that such a tree existed. I had him come to the site, and he made sure that the tree ended up being listed in the BC Big Tree Registry. The registry, now administered by UBC, documents the largest individual of all native tree species in the province. There was no logical or scientific explanation for a red alder of this girth. In all my years of working on the coast of BC, I have never even seen a red alder approaching one metre in diameter. The tree's story will never be known, but in my mind, it must have had a strong spiritual connection with the Haida people of the village.

Late in the summer, Brent, Rod, Tyler, and I travelled to Tofino on the west coast of Vancouver Island. We were going to be doing a helicopter survey of Flores Island, which is a small island (155 square kilometres) located in

Clayoquot Sound, northwest of Tofino. On the southeast corner of the island is where the Ahousat nation (part of the Nuu-Chah-Nulth First Nation [Nootka]) is located. When we arrived the weather was fine, but the next morning we were greeted by a torrential rainstorm. The helicopter was grounded due to high winds, so we hunkered down in the motel, playing cards. The bad weather continued into the next day, so more cards and glasses of scotch ensued. Finally, on the third day of the storm, we gave up and headed back to Victoria. We eventually completed our work in November when it was cold and rainy, but we were able to fly to our sample sites in the absence of lashing winds.

Later that same summer, we were hired by the MOE to collect vegetation data as part of the SEI (Sensitive Ecosystem Inventory) program being done to assess the sensitive ecosystems in the CRD (Capital Regional District). Sensitive ecosystems include Garry oak woodlands, wildflower meadows, old forests (greater than 100 years), wetlands, coastal bluffs, seasonally flooded fields, and sparsely vegetated lands (e.g. sand spits, rock outcrops). Of these, the Garry oak woodlands were and still are under severe threats from development and the encroachment of invasive weed species.

We started work by pre-typing polygons and labelling the various types of ecosystems on aerial photos. Then, we ventured out into the region to field check as many polygons as possible on the ground. Sadly, we couldn't find a single Garry oak site that was in a pristine, natural condition. Everyone of them was compromised by the presence of weedy species.

On the southern portion of Vancouver Island, there are three dominant invasive weed species that all choke out native species. They are Himalayan blackberry, Scotch broom, and English ivy. Himalayan blackberry forms impenetrable thickets in disturbed areas, in ditches, and field edges. It has vicious thorns but does produce buckets of delicious fruit. Scotch broom was brought to Sooke (west of Victoria on the west coast) by a naval captain in 1850. In the 145 years from its introduction, it has spread all the way up the east side of Vancouver Island and even established on the lower mainland around Vancouver. It spreads quickly on disturbed soils, and with a seed pod that explodes when ripe, it can expel seeds into new areas. The foliage is poisonous to humans and animals, and the seeds stay viable in the soil for many years. English ivy wraps its tendrils around trees and throughout the Victoria area. It chokes trees to death.

I had an old Douglas-fir tree in my backyard, and the trunk was covered with ivy. Some of the ivy branches were more than 10 centimetres in diameter. Garden centres on the coast sell varieties of the ivy and broom, and customers buy these 'weeds' without realizing the problems they will have trying to stop them spreading in their gardens.

On one occasion, Tyler and I had ventured into the GVWD (Greater Victoria Water District), where private vehicles are not allowed without previous authorization. We had approval to check on some SEI wetlands that we had mapped and were working off poor quality black and white aerial photos. Finding our sample sites was proving difficult due to a labyrinth of old logging roads. Since the mid-1950s, the GVWD had supplemented costs of supplying water to greater Victoria by allowing logging on selected areas. The logging was always done carefully to not jeopardize the quality of drinking water. In 1994 a judge ruled that the water district had been logging those lands illegally. Subsequently, logging was phased out.

Tyler and I figured we were close to where one of our subject wetlands had to be, so we parked the truck and walked. We were following an overgrown road and eventually found ourselves in an old, logged opening. The opening had young Douglas-fir trees about three years old and three metres tall. Tyler looked at the aerial photo again and said, "I think the wetland is about 200 metres south of us." He showed me the photo, and I agreed, so off we went. Within ten minutes of walking with my head down as I studied the image, I looked up, and to my surprise there was a vigorous, tall cannabis plant right in front of me. I was stunned. As I looked around, I realized that there were many other plants, all in large containers.

I laughed, "Tyler, do you see what this is"? I looked around to see Tyler, with a big grin on his face, busy stuffing leaves into his backpack. I grabbed a handful of the pot leaves then had a thought. "Tyler, we better get out of here. Someone went to a lot of trouble to get all this gear in here and grow these beauties."

He said, "You're right; someone could be watching us right now." With that, we turned tail and headed back to our truck. We kept looking behind us, half expecting to see the cannabis cultivator coming after us with a rifle in his hands. In those days, hidden pot plants flourished on Vancouver Island, and it wasn't unusual for a grower to protect his crop with booby traps.

In all my years of doing fieldwork, it was the only time I found a pot plantation. One of my good friends was keen on finding the plantation, and I showed him on the aerial photo where I thought it was. He tried a few times but was never able to track it down.

24

THE GOLDEN SPRUCE AND A DOOMED FIELD TRIP

One of the projects we undertook for MacBlo in 1996 was for the upper Yakoun River valley on Graham Island. Not far from Port Clements on the banks of the river was the fabled Golden Spruce tree, a magnificent Sitka spruce which was in the order of 300 years old. When our crew was staying in Port Clements, we took the opportunity to view the tree and marveled at its striking appearance. A rare genetic mutation had resulted in the production of beautiful golden needles - in the sunshine, they appeared to glow with an inner light.

Tragically, in January 1997, a troubled, unemployed forest engineer by the name of Grant Hadwin chose to inflict lethal damage to the glorious tree by cutting into its sturdy trunk to the core with a chainsaw, leaving it unable to withstand strong winds[13] (see Footnotes).

In the middle of winter, he must have waded across a channel of the river, hauling the heavy implement of destruction with him. He then sent a letter of explanation to various organizations including Greenpeace, media channels, and the Haida Nation, stating that he reluctantly killed this majestic tree in rage and protest over "... university-trained professionals and their extremist supporters, whose ideas, ethics, denials appear to be responsible for most of the abominations towards plant life on this planet."

Just a few days after his act of desecration, the great tree fell, and the people of the region were overcome with grief and guilt at not having better protected it. Locals gathered in anguish, and memorial services were held. It was as if a beloved elder had been murdered. Shortly after, Hadwin was arrested by the RCMP, charged with the crime, ordered to appear in court, and released on bail. He was apparently last seen paddling his kayak north from Prince Rupert toward Alaska. He never appeared in court.

Later that year, the remains of his kayak were discovered, but his body was never located. It was never determined if he was murdered or if Mother Nature wreaked vengeance on him. Maybe he faked his own death and is still living in the wilderness somewhere.

Prior to its demise, cuttings of the Golden Spruce were grafted onto average Sitka spruce trees and produced saplings. Some saplings survived, including one planted in 2001 in Port Clements to signify a new beginning.

Strangely, I knew Grant Hadwin. I had gone to high school with him in West Vancouver, and we graduated together in 1967. I remember him as a loner with a strange sense of humor. I always wonder what had led him to become so angry with the logging industry and the so-called 'academic elite.'

In September of that year, Shearwater started work on a biophysical survey, including assessing the wildlife habitat of the Itcha and Ilgachuz Ranges. The crew included Brent, Tyler, Rod, Michelle Burfitt, a terrain mapper we hired that year, Laurie Hamilton, a GIS technician and part-time field worker, independent wildlife biologists Stefan Himmer and Damian Power, and I. When I enquired about securing camp accommodations near Puntzi Mountain, located about 60 kilometres south of the study area, we were told that the camp was full, and no rooms would be available. We were left with the only option of staying at Alexis Creek, on Highway 20, in a rundown motel. The only place to eat in the town was the Cook Shack Restaurant. On the side exterior of the building was a large air vent with an ominous grease stain running down to the ground. We suitably called the place The Grease Shack.

We were a long way from our study area and had to drive about 150 kilometres from Alexis Creek before we even could start collecting field data. When we finally were able to get into the mountains, the weather turned cold. Within a few days, we were encountering temperatures hovering around freezing with accumulating snow. Doing our plots became harder and harder, and eventually, we were forced to abandon the sampling program and return home to Victoria.

25 CAMPING IN THE MOUNTAINS

In 1997, we continued the TEM program for MacBlo. In 1999 MacBlo was bought by Weyerhaeuser Canada Ltd. Between 1995 and 2002, we completed 26 mapping projects up and down the coast of BC. The work employed up to 18 people at one time and made Shearwater a highly respected TEM company. Other people who came and went from Shearwater included: data processor and part-time fieldwork Alasdair MacLennan, GIS whiz Anthony Herman, GIS Tech and part-time field worker Shawn Munro, GIS Tech and part-time field worker Peter Kasianchuk, data processor and field worker James Wadsworth, mapper and part-time field worker Roberta Hasselgrave, GIS operator Rosemarie Laird, Rick Marchant (Rick did all of our aerial photo mapping transfer into digital files), data processor and receptionist Cheryl Buckley, and data processor and receptionist Mary Ann Charlie.

That summer, we launched a program to complete the ill-fated field trip of 1996 to do the biophysical inventory and mapping of the Itcha and Ilagachuz Ranges. Rod, Brent, Colin, Michelle, and I made the long drive from Victoria to Anahim Lake. Anahim Lake is a small unincorporated town located about 220 kilometres west of Williams Lake on Highway 20. The population in 2018 was 140. Stefan and Damian met us in Anahim Lake, and we were joined by Jenifer Penny, a botanist with the BC Conservation Data Centre. Jenifer, who had taken time off from work to fill the role of vegetation ecologist, flew into Anahim Lake, and I went to pick her up at the small gravel landing strip. The first thing she said when I met her was, "I forgot my boots." Uh oh, I thought. We might have trouble finding boots in Anahim Lake. We scoured the town and eventually found her a set of high-top, athletic shoes that had to suffice.

Our center of fieldwork operations was a BC Parks cabin which we had obtained permission to use near Itcha Lake in the Itcha Range. The cabin was used as a kitchen, and we all slept in tents amidst the open lodgepole pine forest. One innovation for this project, that was a new wrinkle, was the retaining of a cook to prepare our meals for the time we were camping. Her name was Debbie, and she was the wife of a local rancher. The self-preparation of meals when doing camping field trips is odious and challenging to say the least, and not something that we wanted to do at the end of our long workdays. Having Debbie help us with meals was a hugely successful and popular initiative.

The cabin must have been built by hobbits as the entrance door was quite low. It was lower than two metres because I bashed my head on the upper sill numerous times when departing the cabin. It got to the point where the crew sitting around the campfire would bet to see if I would smack my head every time I came out of the cabin.

One day when Colin, Damian, and Brent were working in the alpine, they saw a grey wolf watching them from a distance. The wolf, which followed them for some time, would sit and watch them work. It never came closer or threatened them; it just seemed genuinely curious about what they were doing.

Late in the summer, we travelled to Fort Nelson to work on the Snake-Sahteneh biophysical mapping project. Our crew consisted of Rod, Brent, Tyler (he had come back to Shearwater that year), Ron, Michelle, Ray Demarchi, a retired BC government wildlife biologist, Don, Ray's son and a wildlife biologist, and I. As part of the project, we were tasked with producing ecosystem maps that would be used to assess habitat for deer, Rocky Mountain elk, woodland caribou, moose, and bison. The study area was also home to grizzlies and black bears.

Rod and I had been dropped off at one site and were completing an ecosystem plot in a trembling aspen stand. We had noticed a game trail through the middle of our plot, and while we were doing our work, we were surprised by a large black bear approaching us. Our heart rates increased dramatically, and we quickly took our bear spray canisters from their holsters. The bear, who was busy munching on low-bush cranberries, kept moving towards us. At one point, he was within ten metres of us and just kept eating and walking past us, all the while keeping an eye on us. He never exhibited any aggressive behaviour and didn't appear to be bothered by us. Eventually, he kept on going about his business and moved away. Rod said that he had enjoyed the adrenaline rush associated with the close encounter, but I didn't share those feelings. It was just another case of meeting wildlife up close and personal.

In all my encounters with black bears, it seemed that they all exhibited different kinds of behaviour. Some were grumpy (they might have had a sore tooth or paw), some couldn't have cared less about us, some were aggressive and noisy, some were just curious, and some were being protective of young. You just never knew what kind of bear would be in front of you. The reality was that we were intruders in their territory, so we always respected the bears that we came across.

Wood bison historically ranged over the study area, although numbers in 1997 were thought to be low. In 1998 the CWS (Canadian Wildlife Service) conducted a release program in the Northwest Territories and the Liard River area (north of Fort Nelson). During the release program, 17 of the bison dispersed southwards and established a functioning herd in the Fort Nelson area.

26 CHRIS CZAJKOWSKI AND WHITEBARK PINE

The Charlotte Alplands was another biophysical project that we undertook for the MOE in Williams Lake in 1998. Our crew for this project was Damian, Stefan, Rod, Brent, Ron, Michelle, and me. We were also accompanied by Stefan's dog, Ochre, a young yellow Labrador retriever. Ochre was comfortable flying in the helicopter and thus accompanied Stefan during field days. He had trained her to dive down into water up to three metres deep to retrieve her favorite rocks.

We stayed at cabins on Whitton Lake owned by Chris Czajkowski. She called her home Nuk Tessli. Chris was and still is a wilderness, off-the-grid dweller. She was born in 1947 in England and emigrated to Canada in 1979[14] (see Footnotes). At some point she decided to live in the wilderness of the west Chilcotin close to the Coast Mountains.

Her first home was built on the Atnarko River with no access except on foot. In 1988 she moved to the location on Whitton Lake, with a hike of 32 kilometres out to civilization. Single-handedly, she constructed her first cabin by hand. She did all the work herself – felling and peeling trees (mostly lodgepole pine, using specimens that were straight and sound, since the whitebark pine were all gnarly, twisted trees that had been shaped by the forces of nature), erecting the support posts, building windows and a door, and then applied a corrugated steel roof. It took her two months (starting on June 30) to have the cabin in a suitable state to afford her protection from the wind and rain. Her account of the building of her cabin and her life in the wilderness is documented in Diary of a Wilderness Dweller. As of the writing of this book, she has published ten books[14] (see Footnotes).

On November 14, she secured the cabin and left for the winter, planning to come back in the spring. When she returned on March 2, 1989, there was over one metre of snow around the cabin site. She continued to work on the cabin from March into April, then went tree planting for two months. I don't think the tree planting was an enjoyable vocation for her, having to deal with the rough camp life and constantly being surrounded by the charred stumps and destroyed landscape of cut blocks. But it gave her the income she needed to continue her nomadic lifestyle. When she came back at the end of June, she set about planning for a second cabin that she started work on in July. By November 21, 1989, she had the second cabin almost finished.

In 1988, Chris had a close encounter of the worst kind with a large, male black bear. On her way out of Nuk Tessli, she and her dog (Lonesome) took shelter in an old trapper's cabin. She had checked out the cabin before and had earlier seen damage inflicted on the structure by a marauding bear. During the night, she was visited by the intruder.

She was awakened by a loud snarl and then heard nothing. Good, she thought. Lonesome had heard the bruin, had barked and scared him away. Dawn was approaching, but it was still dark out. She heard a snort and the bear's large snout appeared in the open window. She couldn't locate her glasses or find the zipper on her sleeping bag. The next thing she knew, the bear was standing in the doorway a few metres away. She had been reading Bear Attacks by Stephen Herrero (1985) and knew from his accounts that black bears could be just as dangerous as grizzlies. Not sure what to do, she grabbed an old axe lying nearby and began to whack an empty fuel drum. The bear charged at her, with its ears flattened, snarling viciously at her. Then the bear backed off, turned, and charged again. Still, she banged the drum as hard as she could. Finally, the bear slowly backed off, turned around and left. He did not run away since that would be losing face but took his time, every so often turning his head to glare at her.

Chris was lucky that the bear was not more aggressive. I had bought a copy of Bear Attacks and remembered reading it to my crew one time around a campfire. Some of the stories made my blood run cold. One account was particularly gruesome. A young geologist was working alone in the Yukon and had been dropped off by helicopter. Her plan was to collect samples for an hour or two then call the helicopter on her hand-held radio when she was ready for a pickup. Not long after she started walking through the bush, she noticed a large black bear at a distance. She yelled at the bear, but it was not bothered at all. As she continued, the bear began to close the distance on her and eventually attacked. She had no weapon and fought the bear, but being of small stature, she stood no chance. As the bear mauled her, she managed to call the chopper pilot on her radio. When the pilot returned and found her, she had been severely injured. The bear had peeled the flesh from her arms like two bananas.

She survived the horrific damage, although I believe she lost both of her arms. I agree with Chris – black bear attacks are far more common (partly due to their wider range and proliferation of blacks versus grizzlies) in Canada than people realize. In the last 40 years in North America, the number of fatal attacks, comparing black bears and grizzly bears (including brown bears), are not that different (50 and 42 respectively). However, it should be noted that an individual has a higher chance of dying in a car accident, from a bee sting, or being hit by lightning. We never finished reading the bear attack book because some of the attack accounts were even more shocking. I have always thought I have been lucky in all my years to never have been attacked, and I have had lots of encounters.

The bottom line is that when you are in bear country, making lots of noise is the best policy and keeping your eyes and ears open is paramount.

I think Chris patterned her pioneer life after Ralph Edwards, who, at the age of 21 in 1913, was granted a homestead on the Atnarko River near Lonesome Lake. He lived there for many years, being totally self-sufficient save for an occasional packhorse trip of two days to civilization to refill supplies. The lake was home to a gaggle of trumpeter swans, reduced to near extinction by the beginning of the 20th century. Ralph took a personal interest in them, providing food to them in the winter. In 1925 the federal government entered into a formal agreement with him to continue to provide food to the swans in the winter, which he did for many years. He was the subject of the book Crusoe of Lonesome Lake by Leland Stowe in 1957, and in 1972 he was made a member of the Order of Canada.

Similarly, as in the previous year at Itcha Lake, I devised an ingenious method for us to have dinners prepared. Jim and Mary ran a lodge on Nimpo Lake, and we met them when we stayed there. We arranged for Mary to cook us dinners and Jim to fly them from Nimpo Lake, a 20-minute flight, to Nuk Tessli. Every night at about the same time, we would eagerly await the drone of Jim's Cessna as he came in for a landing. We were fed very well, and it was always exciting to see what feast Mary had prepared for us. She was an amazing cook; her apple pie was the best.

The Charlotte Alplands are home to one of BC's most extensive populations of whitebark pine[15] (see Footnotes and Appendix). This pine species is typically found in parts of BC where the climate changes from coastal to interior, thus being tied to the transitional nature of ecosystems that are present. The pine trees have a unique relationship with the Clark's nutcracker, a member of the Corvid family. The most important food resources for the birds are the seeds of whitebark pine, which are particularly large. Clark's nutcracker is the primary seed disperser for whitebark pine, which is in decline throughout its range due to infection by white pine blister rust (*Cronartium ribicola*), widespread outbreaks of mountain pine beetle, and the long-term effects of fire suppression. If the whitebark pine decline continues into extinction, the nutcrackers will lose an important source of food and may no longer be seen in areas where the tree is the primary source of seed (mountainous areas of BC and the Rockies in Alberta).

The Charlotte Alplands eventually was preserved forever in its spectacular state by being designated a Provincially Protected Area. No roads or resource extraction would ever desecrate her amazing mountain peaks or sublime subalpine meadows.

14 BEARS AND THE RVS

In 1999, Ron, Colin, Rod, Shawn, Michelle, and I drove to Port McNeil on northern Vancouver Island to complete TEM work on MacBlo's TFL land northwest of Port McNeil. The area is one of the wettest places on Vancouver Island, with an average annual precipitation of more than 1,800 millimetres, of which the majority falls as rain. For much of the year, it is damp and gloomy. We used to call it the 'armpit' of the island.

Port McNeil has a long history of logging, and much of the area we were surveying had been previously logged. Logged areas regenerate quickly with berry-producing species such as red elderberry, salmonberry, Alaskan blueberry, and American red raspberry proliferating. Because of the wealth of food available, there are a relatively high number of black bears present. In one day of working along the logging roads, Colin and I counted 14 bears. Amazing to think you could see that many bears in one day, but we did.

The continuous array of shrub species provides plenty of food for another native animal, the black-tailed deer, a favorite prey species for cougars. It is said that Vancouver Island has the highest density of cougars in North America, due primarily to the abundance of deer. Unfortunately, with that many cougars around, there are occasionally incidents of them stalking people and pets, even taking small dogs in front of owners. I had never seen a cougar until one day on the way home from the bush when Colin was driving. I was nodding off in the front seat when he suddenly yelled. I thought at first that we were going to be dodging a large logging truck, but I opened my eyes to see the tail of a large cat disappear into the roadside vegetation. To this day, it has been my only opportunity to see a cougar, and I ultimately only saw the back half of it.

There was no helicopter flying requirement for this job, so we had Milo, our family dog, a large black Labrador retriever with us every day. When I did fieldwork that was all ground-based, I would bring him along as a companion and guard dog for bears. He loved being out with us, keeping us company, watching for bears or other wildlife, and bugging us to pick up sticks and throw them for him. He relished eating salmonberries and would pluck ripe ones from the stems and gobble them down. I had heard stories of people's dogs bringing bears back towards the owner if a bear was chasing the dog, but this never happened in all the time I had Milo in the bush with us. One day in the Port McNeil work area, he did chase off down the road and around a corner, after a small black bear. But he eventually came back to us without a bruin chasing him. The poor guy was always pestered by mosquitoes. They would invariably go for his nose where there was very little hair.

Late in the field season we were asked to do more TEM work for MacBlo in the Sarita Creek drainage on the west coast of Vancouver Island. The closest location for a motel or hotel was Bamfield, but when I made phone calls, there were no rooms available anywhere. As an alternative, and this was the first and only time we did this, we rented two large recreational vehicles (RV), each able to sleep three people. Our crew consisted of Rod, Michelle, Colin, Ron, Shawn, and I. Milo came on this trip too. We headed up the Island Highway and down a logging road towards Bamfield, on the west coast. As we hurtled up the highway, I felt like I was driving a large bread box on wheels. Once we left the asphalt and got onto the logging road, the fun really began. The large RV bounced crazily over the bumps and washboard road surface. I could hear cupboard doors in the vehicle flying open and dishes, pots and pans, and cutlery flying out of cupboards. It was as if I was Desi Arnaz trying to control the trailer in the 1954 movie The Long, Long Trailer[16] (see Footnotes). Lucille Ball was his newlywed wife in the movie, and on their way to Nevada, she collected a huge number of rocks with the intention of using them to decorate the yard of their new home. At one point, Desi discovered the rocks, which had caused the trailer to become unbalanced and difficult to pull and orders her to dispose of them. Lucy disobeyed him and stashed the rocks as they prepared to climb over the mountains. As they reached the mountain pass at over 2,000 metres, the trailer became hard to control on the narrow winding road. The rocks which Lucy had stashed came bouncing out, and the trailer ended up going off the road and becoming stuck. Our journey was nowhere near that bad, and eventually, we arrived at our destination, the Pachena Bay campsite, with all our kitchen wares strewn across the RV's floor.

The campground, which is situated right on Pachena Bay, is part of the commitment of the Huu-ay-aht First Nation to sustainable economic development and further their vision of working together for a healthy and prosperous community. Pachena Bay, approximately three kilometres long and one kilometre wide, faces the open Pacific Ocean to the southwest. At the head of the bay, directly in front of the campground, is a magnificent, natural sandy beach approximately 1,200 metres long. At low tide, the beach is nearly 200 metres wide, ideal for beachcombing, relaxing, jogging, and for observing the abundant marine wildlife. We enjoyed our stay there, and the RVs, although somewhat cramped, served us well.

One field day, Colin was driving the company SUV, a red Dodge Durango which did not have great road clearance, up the main road from Pachina Bay and onto a side logging road up in the hills. We climbed steadily up and completed a few ecosystems plots along the way. Just before lunchtime, the road had started to look sketchy, but Colin, who was 20 at the time, was doing okay, and I figured the 4 x 4 experience would be good for him.

Eventually, we were driving in a heavily rutted, steep section of road with poor traction. Before I knew it, the vehicle slid sideways, and Colin had gotten us very stuck. When we were on backwoods roads, we never knew what we might encounter. In front of you could be a washout, a creek gully, a sinkhole, trees or boulders on the road, brush overgrowing the road, or a dead end. These days if a road looks at all suspect, I usually stop and walk or turn around and look for a better route. On this day, I should have warned Colin earlier before we got stuck, but I didn't. We tried for about an hour to shore up the ruts with pieces of wood and rocks, but we were badly high-centered.

After two hours, we realized that our situation was futile. We would not be driving the Durango back down the hill that day. Milo was with us, and he was unaware of our predicament as we started walking back the way we had come. He was happy to be free of the bouncing in the back of the vehicle. It took us about two hours to walk down to the main gravel road where we sat and hitch-hiked. It was about 3 pm. The only problem was cars or trucks going our way were few and far between, and no one would stop.

By around 6 pm, our situation was not looking good, dusk was approaching, and we still had not gotten a ride. Finally, at close to 7 pm, salvation arrived in the form of Rod and Michelle. Since we hadn't arrived at camp by 6:15 (that would have been unusually late for us to return from the field), they decided that something must have happened to us. They knew the general area where we had planned to work and headed there. They were as relieved as we were to find us by the side of the road with a small fire going as darkness descended. The next day we found a tow truck operator in Bamfield, and he took us back to the Durango. We had her free in about 20 minutes.

At the end of 2000, Michelle left us to travel to England with her husband to embark on a new adventure. To replace her terrain mapping skills, we hired Dr. David Huntley, an accomplished surficial geologist (same as terrain mapper). David grew up in England and came to BC in 1987. He was as eccentric a fellow as I have ever met, but I have always believed that our type of work attracts eccentric scientists. Roaming around in the woods for most of one's life would not be comfortable or interesting to most of the general population. When David started working with us, the rest of the crew was a bit wary of him, but we soon came to be good field partners with him. He formed a special bond with Rod, and for the next few years, they were a constant field team.

We also hired my old mentor, Jim van Barneveld. Jim had taken early retirement from the BC civil service, but he still had some good years left in him, and we needed another vegetation ecologist. I loved Jim dearly, but he had been in the government for so long he never really adapted to the world of consulting. In consulting, the old tenet 'time is money' is absolutely true. Often in the government world, accountability is not a critical component of work productivity. In consulting, accountability drives the business, both for small companies and clients (who pay the bills). Jim only worked for Shearwater for two years. Still, during that time, he managed to lose a $1,200 radio, locked the keys inside a truck when the crew was in the middle of nowhere, forcing his partner to break the truck window to get the keys, messed up countless computer files, and allowed map units typed onto aerial photos to wash off in the rain. But I forgave him for all those indiscretions. It was definitely strange being his boss after he had been my boss for ten years at the start of my career.

28

SAVARY ISLAND SAGA AND 911

Early that summer I was contacted by Roger Sahlin, an American businessman living in Bellingham, Washington, requesting us to complete ecosystem sampling and mapping for a portion of Savary Island. Savary Island is located just west of Powell River, BC, in Georgia Strait. The island is small, only about eight km long. Access is by boat or float plane. Ron and I took the water taxi from Lund, about a 20-minute trip. There are no services on the island, which means no general store, no post office, no gas station, and no emergency responders. The people who live there full-time (there are many seasonal residents) like their privacy. They enjoy the solitude, lack of amenities and freedom from rules and regulations. Following deglaciation, about 10,000 years ago, much of the island had been covered in sand dunes.

The property that Roger wanted us to assess was DL1375. It had been in his family for many years; his dad logged part of it in the 1950s. It was a large tract of land, approximately 54 hectares, in the middle of Savary Island. He had a partner who was an equal owner, although I never met the partner or even heard Roger mention his name. Roger and the partner never formally decided who would own what portions of the property, which ultimately turned out to be a mistake.

Ron and I walked to the project area from the dock where the water taxi had left us. We completed our sampling in one day, caught the water taxi back to Lund, and headed back to Victoria. When we finished our report, Roger was quite happy with the results, and I assumed that would be the end of work with him. Little did I know how wrong I would be.

Over the next 12 years, I would go back to the island another four times and become embroiled in a lengthy legal battle over property rights that would last three years. In 2004 I flew to the island in a float plane with Roger. He wanted to show me what his plans were for building recreational cottages for his family members. We landed in the waters off the south shore and walked up a trail through the mixed western hemlock and veteran Douglas-fir trees. We perused the property for around two hours and then headed back to the plane. Unbeknownst to the pilot and us, the tide had switched and was heading out. No one had checked the tide charts for that morning. So, there was the float plane stranded high and dry on the beach, and it was only about noon. We knew that we were going to be stuck for a while, so we spent time relaxing on the logs at the top of the beach, eating lunch, taking a nap, and beachcombing. Finally, at around 6 pm, the tide had risen enough to allow us to float the plane free and head for home.

In 2006, Roger contacted me and let me know that his invisible partner had created a nightmare for him. Without consulting Roger or telling him, the partner had sold his 'share' of the property to the BC Nature Trust. The Nature Trust is a non-profit conservation organization dedicated to preserving key natural areas of BC in perpetuity. Often, property owners would donate pristine, undisturbed areas to them. The problem for Roger was that the property areas had never been allocated between him and the partner. Four subdivision lots were identified within DL1375, but now a fight began between Roger and the Nature Trust as to who would get what. I went back to the island for a third time, with instructions from Roger to complete a plant inventory along the sensitive portions of the property, which included the sandy beach, the associated foreshore, and the tops of inland sand dunes.

In the court proceedings, I was asked to give a deposition regarding my ecological surveys and veracity of various components. Finally, in 2010 a judge decided on an equitable solution to the disagreement. Roger would be awarded the two middle lots (lots 2 and 3), while the Nature Trust would receive the west and east lots (lots 1 and 4). Nature Trust's lots had public access (on the west and east extremities); Roger's lots were the ones that would allow him and his family to build their small cabins away from the protected areas. In addition, the foreshore and sandy beach areas were set aside as a conservation area to protect an endangered plant.

Following the court verdict, Roger had me go back to the island in 2012 to complete a formal EIA (Environmental Impact Assessment as per the Canadian Environmental Assessment Act) that had been requested by the BC Ministry of Transportation and Infrastructure. The assessment was directed to be done so that minor improvements could be made to the access road running through Savary Island, as well as allowing access to the Sahlin family building sites. I spent two days staying on the island while doing the work, which included documenting plants and animals on Roger's property. I submitted the report to Roger but never heard back from him.

The following year I was asked to travel once again to the island to help with a survey to determine the population numbers of the endangered plant that grew on the beach front. The plant in question was contorted-pod evening-primrose (*Camissonia contorta*), known from only six locations in southern BC, including Savary Island. It was such a small, obscure plant I had never even documented it in my two plant surveys. I was accompanied to the beach foreshore by Matt Fairbarns, a species at risk and rare plant expert, and within two hours of searching, we found more than 100 individuals of the diminutive plant. It was only when it was flowering, early in the growing season, that it could be seen. Matt was excited about our findings, and I learned something new. It was a common thread of all my fieldwork – no matter where I went, I would always see something new and different. I would constantly be stimulated by some fascinating aspects of plants, animals, or the landscape I was working in.

I never did hear again from Roger Sahlin. The bill for the EIA had never been paid (I was working for an environmental company when I did the work). Then out of the blue in 2018, I was contacted by an associate of Roger's. I found out that Roger had declared bankruptcy, but the EIA bill got paid.

A significant event of 2001 was the BC general election. The NDP (owners of the FRBC funding) lost the election and were annihilated in the process. They had been in power since 1991, but in the 2001 election, they only won two seats. The Liberals delivered one of the worst election defeats in Canadian political history, taking 77 of 79 seats. Two scandals (the Fast Ferries fiasco and a bribery incident involving Premier Glen Clark) ultimately sealed the fate of the NDP. Within a few months of taking office, the Liberal government scrapped the FRBC program and replaced it with the Forest Investment Account. This initiative was oriented toward funding projects that would result in increases to the AAC (Annual Allowable Cut) for BC. With the demise of FRBC, funding for TEM projects dried up.

September 11, 2001, found Rod, David, Ron, and I set to start work on the ecosystem mapping of TFL 33. The TFL, which is situated north of Sicamous, BC bordering on Shuswap Lake, was managed by Federated Co-Operatives Ltd. Today it is operated by Canoe Forest Products Ltd. We had risen from bed that morning and were eating breakfast when we flipped on the TV. I heard Ron say, "Oh my God!!!" and audibly gasp. As he said that, I swivelled around in my chair to look at the image on the TV, and there were the North and South Towers of the World Trade Center with smoke pouring from them. I said to Ron, "Holy shit. What happened"? Ron said he didn't know, and we watched transfixed as the horror unfolded in front of us. The crashes had happened respectively at 8:46 and 9:02 Eastern Time in New York. As the television commentator revealed the magnitude of the terror attack, we sat mutely staring at the images coming to us. We quickly let Rod and David know what had happened, and in minutes we were all in the same room staring at the screen. It soon became evident that a terror organization had launched a horrendous attack by hijacking four planes and using them as weapons. In all, 2,977 people died in the attacks. The USA and the rest of the world were put on alert to the potential for more attacks. We were stunned and in shock.

I recall that we didn't go to work that day. We were scheduled to conduct much of our work using a helicopter based out of Salmon Arm. Due to the terror attack, all flying across North American was grounded for two or three days. We never did use a helicopter to fly into the TFL. We did go to work the next day, and we used trucks to complete as much ecosystem sampling as possible. The mood while we were working was sombre and quiet.

I have very few memories of the actual fieldwork, although I have one specific memory regarding a pile of bear scat that Ron and I encountered. We were walking down an old, inactive logging road when we came upon what was probably one of the largest piles of bear dung that I had ever seen. You can often judge the size of a bear by the diameter or bore of his scat. In some cases, especially if the bear has been eating fruit, the scat will be loose and lacking form. This one was like a small log and more than six centimetres in diameter. It so impressed me that I put a small ruler beside it and took a picture. From the size of it, I was pretty sure that it had to be a grizzly, a considerably large one. Ron and I kept our eyes and ears open as we continued down the road, and after an hour, we were happy to be back at our truck.

20

PREDICTIVE ECOSYSTEM MAPPING AND ALONE AGAIN

In 2002 we won a contract to conduct an audit on a PEM (Predictive Ecosystem Mapping) project for the Merritt, TSA (Timber Supply Area). Predictive Ecosystem Mapping was a government invention designed to reduce the cost of ecosystem mapping and theoretically replace, or at least supplement, traditional TEM projects. TEM projects were costing in the order of $4.00 per hectare, which meant for large projects, the bill might exceed $500,000. I'm not sure how it started or who was responsible for initiating the PEM program, but I was not entirely convinced of its actual usefulness. It involved combining existing resource information such as forest cover mapping, terrain or soil maps, and topographic maps and using a computer to generate PEM polygons. A polygon was just a piece of any landscape defined or delineated by a closed polygonal line. With TEM projects, once the landscape was mapped, the polygons with their associated attributes could be used to develop endless interpretations, such as resource productivity and capability [e.g. forestry, recreation, wildlife]. The main problem was that the PEM mapping was only as good as the information that went in. Typically, forest cover maps mainly were based on history (logging, forest fires, or other disturbances) and not on ecological traits or environmental features.

Rod and I spent two weeks covering the TSA, a huge area (1.13 million hectares) from south of Kamloops almost to the American border. Our polygons for checking were pre-selected, and so we had to conduct hunts to find them. Often our days were spent just figuring out how to get to the polygons, with many times roads becoming impassable due to washouts or overgrowth of shrubs and trees.

Inevitably, when Rod and I checked the PEM polygons, the mapping was erroneous. Not only erroneous, but way out. It became apparent that the computer-generated mapping (that was still expensive) was not worth the money that the government paid for it and, unfortunately, did not provide a viable solution to the cost of TEM.

Later that year, Shearwater Mapping Ltd. closed its doors. With the loss of government funding for TEM work, most forest companies did not take on the cost of doing the work. Our permanent employees were given severance packages as most of them had been with us for six or seven years. All of them, bar none, were successful in finding gainful employment with other companies.

I was alone again and registered another proprietorship, Clement Consulting. Over the next few years, Ron, Rod, and David continued working with me on numerous TEM projects and other environmental assessment work.

SECTION 5

Working Independently Again

When we shut down Shearwater, I continued working in the consulting business as a proprietor. My career spanned 32 years by now, and I was in demand as an experienced vegetation ecologist. Somehow, the projects became less exciting over time, or maybe I was becoming jaded as I got older. Whatever the reason, there were fewer comic events in my work life from that point on. I also did more jobs involving linear environmental impacts, such as power lines, highway improvements, or pipelines. In many cases, those types of jobs are not glamorous and are actually monotonous. Such is the nature of the business. In the immortal words of Bob Dylan,

"Well, it may be the Devil

And it may be the Lord

But you're gonna have to serve somebody."

(from Slow Train Coming, 1979)

Everyone has to serve someone in the environmental science field, whether you work for government, industry, consulting, First Nations, or non-government agencies. It is all a matter of personal choice, ethics, and paying your bills.

LAST TRIP TO HAIDA GWAI AND THE RAVEN

In 2004, Rod and I travelled to Haida Gwai to do a TEM 'light' project in the Queen Charlotte TSA for Husby Forest Products. TEM light projects were considerably cheaper than traditional TEM, with cost savings being realized by reducing field sampling requirements. They were essentially a hybrid of the PEM and TEM methodology. We were picked up at the Sandspit Airport by a VIH (Vancouver Island Helicopters) chopper and flown to a logging camp in the northern part of Graham Island. The camp was a sprawling cluster of old buildings with cedar plank siding and corrugated steel roofs. Around camp were scattered tame black-tailed deer who seemed as comfortable as household pets. Along with the deer were resident ravens who regularly scavenged handouts from the loggers.

One morning as Rod and I left camp in our dilapidated, old pickup, we noticed a raven flying right beside the truck. Whenever I stayed as a visitor at a logging camp, I was always given a truck that had invariably seen better days. As we followed the logging road out of camp, the bird kept pace with us. I looked over to my left, and there it was, going the same speed as the truck. We drove for about an hour before reaching our first sample site, and when we stopped, the bird perched on a nearby log and watched us intently.

Rod and I finished the plot and then proceeded to the next sample site. The raven followed us again, and like before, he watched us while we worked. The same process was repeated until, at one stop, we pulled out our lunches. I knew why he had followed us and what he was waiting for: handouts. Of course, we obliged him by throwing some sandwich crusts his way. After lunch, once he was satiated, he flew off, probably back to camp.

What an intelligent bird to recognize that he would most assuredly benefit from free food by following workers and their trucks. Just another example of the brilliance of the Corvids. One of the pilots told me that he had seen ravens flying at young deer to drive them off cliffs, thus killing them and providing food.

While Rod and I were at the Husby camp, we were graced with the presence of two MOF personnel who had been sent into the field to check on my ecosystem classification and mapping skills. I don't remember their names, just that they had no idea regarding the ecosystem classification of the area of the Queen Charlottes where we were working. It was ironic that for them to check my work, I had to 'train' them in ecosystem classification first.

31

THE UPPER NICOLA BAND

In 2006, out of the blue, I received a phone call from someone by the name of Gary Arnold. I was initially suspicious since I knew of no one by that name, and he purportedly wanted to hire me. After listening to him talk for a while he explained how he had found me and what he wanted me to do. It turned out that Gary was supervising a Natural Resource Technician program for the Upper Nicola Band at Douglas Lake, BC. Douglas Lake is about 52 kilometres east of Merritt. The landscape in that region is characterized by rolling grasslands, with scattered stands of ponderosa pine and Douglas-fir, as well as black cottonwood watercourses and trembling aspen copses. Gary had somehow met Anne's brother Nobu. I think Gary had befriended Nobu and put him up at his house in Merritt for a few days. I have no idea how my name came up, but during Nobu's interaction with Gary, Gary learned about my experience with ecosystem and environmental resources. So, that is how Gary came to cold call me.

The Upper Nicola Band controls 12,200 hectares within eight reserves, and the Douglas Lake Reserve is the largest at 9,327 hectares. The Band is part of the Okanagan Nation. Gary wanted me to come to Douglas Lake and spend time training three young technicians who had been hired to do environmental work for the band. Two of the technicians were band members; the third was an adopted member. I was not busy at the time and was low on money and work prospects, so I took him up on his offer.

I took the bus to Merritt, where Gary picked me up and took me to his house. He was kind to me and gave me a helping hand when I needed it. I spent a few hours talking with him about how he expected me to assist them. He had a plan to train the three technicians to make them a more useful resource for the band. His idea was for them to become proficient at environmental impact assessments, carry out forestry work, collect ecosystem data, and describe the multitude of environments within the Douglas Lake Reserve.

The next morning, I drove with Gary to the band office at Douglas Lake, where I was introduced to my three students, Christine Saddleman, Casey Holmes Jr. (Casey passed away in 2018), and Kelly Whitecap. I planned to spend five days with the students starting with a basic indoctrination into the various components of environmental assessment, including the description of forest, vegetation, and soil resources.

I started off trying to determine what level of training they had received at NVIT (Nicola Valley Institute of Technology). NVIT was initiated in 1983 as one of two BC Indigenous Public Post-Secondary Institutes. Christine, Casey, and Kelly were all graduates of the Natural Resources program. My three students had completed courses in fire dynamics, soil science, silviculture, field surveys including environmental assessment, and ecology (wildlife, grassland, and forest).

For the first two days, I expanded on information they had already been exposed to. I went through the Canadian System of Soil Classification (Soil Classification Working Group 1998) manual with them, stressing the kinds of soils they would see on the reserve lands. The grassland soils would be quite different from the forest environment soils. I quizzed them on how much they knew about forest and grassland ecosystem classifications. I went over common species of the grassland landscape, the evergreen forests, and the trembling aspen groves. We talked about the role of historical fires in maintaining the expansive grasslands. For the rest of my first week with them, we drove to sites throughout the reserve. We dug soil pits in the grasslands, under trembling aspen forests, and in the ponderosa pine and Douglas-fir stands. We sat around the soil pits and talked about the various features such as soil texture, drainage, the rooting zone, and productivity. And we studied the common plants of those ecosystems including, big sagebrush, common rabbit-brush, bluebunch wheatgrass, and arrow-leaf balsamroot.

We found and examined many of the plants that the Okanagan people had used (and still use today) including chokecherry, saskatoon, and soopolallie. The use of native plants by Indigenous peoples goes back thousands of years, and Christine, Casey, and Kelly were keen to learn more about those plants. Big sage leaves were used to make a tea for treating colds. The leaves were also smudged to cleanse living quarters[17] (see Footnotes). Ponderosa pine needles were woven into intricate baskets, and the seeds were harvested from the large cones and eaten. The list goes on and on. Saskatoon berries, which are today grown commercially in parts of Alberta and Saskatchewan, were used as a laxative, and the stems were turned into arrows and pipes. Black cottonweed fluff from the mature seed pods was used for padding, and the sticky, resinous buds, which appear early in the spring, were used as glue and ointment. The students were all eyes and ears as we hunted for the various plants and then read about the traditional uses.

Soopolallie was a favorite of many First Nations peoples both in the interior and on the coast. The berries were an important trade item with a special use as 'Indian' ice cream. The berries, which contain glucoside-saponin, were whipped into a soapy, frothy foam and then eaten as a refreshing delight. The berries, rich in iron, were also eaten fresh or dried. Snowbrush leaves, a favourite food of deer, were used to make tea for treating tuberculosis. Chokecherry, which is prolific in gullies and on dry slopes, was another favorite berry food. I have seen coyote scats full of berries and seeds in the fall. Interestingly, the large seeds contain amygdalin which, if consumed, is converted to cyanide. Prickly-pear cactus, which has vicious, barbed spines, was harvested, and the stems were peeled, then boiled, roasted, or pit-cooked.

As well as studying the use of the native plants, we also focused on some of the common poisonous species. Upland larkspur, with its beguiling, spectacular blossoms, is a member of the Buttercup family and is toxic to livestock. Many species in this group contain alkaloids and glycosides, which are deadly, although genera such as monkshood, buttercup, clematis, and pasqueflower are used in homeopathy.

I returned home to Victoria after a week in Merritt and then prepared for another session with the three technicians in two weeks. Once again, I took the bus again to Merritt and stayed with Gary for another week of work. The first day I was at his house, I didn't have to teach, so Gary and I discussed his goals for Casey, Christine, and Kelly. He told me that he expected them to develop a reasonable level of competency in such things as regeneration surveys, plant identification, watercourse mapping, pre-disturbance environmental assessment, and wetland surveys. I was pretty sure that we could accomplish all of that with two months of instruction. I also told him that I was adamant that most of the teaching would occur in the field. I wanted them to get their hands dirty and eyes opened to the wonders that abounded in their backyard. In the afternoon, I went hunting for a few plants that I would use to play a teaching game with them.

The next morning when I arrived at the Douglas Lake offices, Christine, Casey, and Kelly were waiting for me and ready to get to work. I told them I had a short plant identification exercise for them. I pulled three old ties from my backpack that I had brought from home. I then asked them to place the ties over their eyes and secure them behind their heads so they couldn't see. Once they had the ties securely in place, I explained to them how this exercise would work. I was going to give them three different plant species, one at a time, and ask them to use their hands and noses to note the differences between them.

The first plant I gave them was a snowbrush with large sticky leaves and a sweet, spicy smell. As they received the specimens, Casey immediately said, "Hey, it's really sticky, and it has an interesting smell." Christine and Kelly took a bit more time, but eventually, they agreed with him. After a minute or two, I told them to remove the blindfolds and let them have a good look at the leaves on the stem, allowing them to become familiar with them. The whole purpose of the exercise was to teach them to use their noses and hands as additional tools for plant detective work. Over the years, I had come to learn that one's nose and hands were valuable aids when identifying plants. Many plants have characteristic odours: yellow cedar foliage smells like cat urine (although I find it quite pleasant and striking), red elderberry leaves have a pungent, smoky smell, and skunk currant is described as having a rank odour. I have often depended on the characteristic smells of certain plants to be sure of their epithets. Other plants such as white bog orchid (has an odour of aromatic, sweet perfume), chocolate lily (has an unpleasant smell of rotting, dead flesh which attracts flies who provide pollination services), and tansy ragwort (has an overwhelming, unpleasant smell) have strong scent signatures that are reliable tags.

I told the three of them to put on the blindfolds again, and I handed another plant to them. This time I gave them a branch of big sagebrush. When the leaves are rubbed between the fingers (which I told them to do) a strong, herb-like aroma is released. The members of the sage group of plants have a characteristic aroma that provides a warm, menthol affect to your olfactory chambers. In addition, the leaves and stems are covered in a layer of fine, silver-grey hairs which gives them a downy feeling like felt. The techs all crushed the leaves as I told them to do and immediately were struck by the strong smell and impact on their noses. They removed the blindfolds then looked at the sage stems. The leaves have three teeth at each end (hence the name *Artemisia tridentata* [three-toothed]) and, as mentioned, are covered in fine, grey hairs. I gave them magnifying glasses so they could see the arrangement of the appressed hairs.

The last plant that I gave them, sight unseen, was common juniper which occurs on dry, rocky sites throughout most of BC. As I handed them the branches, I told them to hold them carefully. The needles of common juniper are sharp and prickly so that grabbing a stem with full hand pressure can be somewhat unpleasant but not overly hurtful the way a rose stem would be. They did so, and Christine immediately said, "I know this one, it's juniper, right?" I told her she was correct. Common juniper has the widest distribution of any woody plant in the world, occurring around the entire world in northern temperate climates. It occurs right across Canada except for the Arctic. The berries, when squeezed, have a strong smell similar to turpentine. In different parts of the world, extracts from the fruit are used in production of alcoholic beverages, including gin and ale.

I collected my ties, and then we talked more about how useful our hands and noses can be to identify plant species. Everyone appreciated the 'hands-on' lesson, and we continued the day doing more plant identification work.

Over the next few weeks, I continued to work with the three students and at the end of my time with them, I was sure that they had gained a deeper appreciation of the resources that abounded within the reserve lands.

Gary asked me if it would be possible to use any of the technicians in future field jobs, and I told him I would keep them in mind. The year after I trained them, I was able to take Casey on a week-long field trip to Wells, BC, to collect environmental information for a proposed gold mine project. He was a reliable, cheerful, and competent field assistant, and I enjoyed working with him.

Casey and I were tasked with completing ecosystem plots at

selected sites within the proposed mine footprint and collecting vegetation samples for heavy metal testing. Certain plants are good indicators of heavy metals in the soils, as they uptake the metals where they accumulate in foliage. While there is a multitude of potential contaminants, the significant heavy metals that were going to be tested for were primarily arsenic, lead, and mercury. The rationale for testing plants for heavy metal uptake was that during gold processing, the metals are common byproducts. The plants would reveal current baseline levels.

For our work, we were required to collect samples from subalpine fir, Monk's-hood lichen (*Hypogymnia physodes*), and red-stemmed feathermoss (*Pleurozium schreberi*). At each sample site, we collected plant material in the following manner. All collected material was placed in paper bags. We took the bagged samples back to our bunkhouse every day, where we opened the bags and placed them on the floor so the plant material could dry. The moss was easy to collect because it was abundant, forming a thick carpet on the forest floor, and we could grab large handfuls of it. For the lichens, which grew on the stems and branches of the subalpine fir, we had to painstakingly pluck them. The subalpine fir samples themselves had to be clipped from the tree. We had to obtain the current year of growth, which was at the end of branches, using a long-handled set of pruning shears. Over five days, we collected enough plant material so that a comprehensive analysis of heavy metal content could be derived from the three plant species. At the end of work, I drove back to Merritt with Casey and said goodbye to him.

Shortly after that project, Gary either quit his job with the Upper Nicola Band or was fired. I never found out what had happened, and I lost contact with him. The band never retained me for any additional work. I had enjoyed the experience of working with Gary, the band's staff, and the technicians. I am sure that they all continued in the resource line of work, and I had opened their eyes to the endless types of tasks to be done and information to be collected.

32 THE A-TEAM

I had previously worked for Roger Pope, a senior environmental scientist with Dillon Consulting Ltd. Dillon, is a smallish company with about 700 employees and operations primarily in Ontario. They had a small office in Richmond, BC and lacked the specialized expertise and experience necessary to do TEM surveys. In 2007 I got a call from Richard asking if I could take on a TEM job for a proposed copper mine at Harper Creek up the North Thompson River valley. He indicated that our crew would include Denny Maynard and experienced wildlife biologist Mark Fraker; Roger always referred to us as the A-Team.

The proposed open-pit mine was planned to produce 70,000 tonnes of copper-gold-silver ore over a 28-year expected life. A scheduled operational start date was 2019, with a projected capital cost of close to a billion dollars. To get to that point, the company (originally Harper Creek Mining Corp. and then Yellowhead Mining Inc.) would have to negotiate their way through a labyrinth of environmental assessment requirements, interaction with First Nations groups, and various federal and provincial government agencies. As of 2020, the mine is still not operational, but other companies have expressed interest, including Anthill Resources Ltd. and Taseko.

So off we went, driving up the Trans Canada Highway, onto the Coquihalla (Highway 5), then following the North Thompson River to Vavenby, 545 kilometres from Vancouver. At Vavenby, we left the highway and headed east up a good gravel road to a small work camp where we would be staying. Denny, Mark, and I had done five projects together going back to 2003, and we worked well as a team. We had good road access through the study area, which allowed us to stop at selected sample sites. As the surficial geologist, Denny would ramble away for a few hundred metres in various directions as I dug a soil pit and compiled my list of plant species. Sometimes I would utilize a road cut which was more efficient than digging a pit in the forest; I could always see a deeper soil profile in a road cut than in a hole dug at an undisturbed location. Mark would roam around and look for obvious signs of wildlife use.

Our work proceeded uneventfully, covering Douglas-fir forests at the lowest elevations, which transitioned into a mixture of subalpine fir and Engelmann spruce at the upper elevations. At the highest extent of the study area, we were working in subalpine parkland (islands of subalpine fir interspersed with rich, verdant subalpine meadows). In the subalpine environment, we found ample evidence of bear and moose, but we never had any close encounters with either.

In a small fen, surrounded by thickets of slide alder and white-flowered rhododendron, I came across a patch of common butterwort, a small, striking plant that is insectivorous. It has a basal rosette of pale leaves, a stem of about 15 centimetres and a striking, disproportionately large, violet-coloured flower. Leaf surfaces contain two types of glands. Stalked glands are tipped with globular, mucilaginous (sticky) secretions and sessile glands that release digestive enzymes. Small insects landing or crawling on the leaf surface become increasingly mired in the sticky, honey-like secretion, which stimulates the production of the enzymes. The prey gradually becomes pooled in digestive juices, and these nutrients, including phosphorus, are gradually absorbed into the leaf. It is also suggested that other nutrient-rich substances, including pollen, are also utilized by the plant. Leaves also produce a bactericide that retards the insects from rotting while they are digested.

Insectivorous and carnivorous plants have always fascinated me, and there are others that have different ways of capturing and digesting insects. Probably the most striking and uncommon of these is the pitcher-plant[18] (see Footnotes). I have only seen it on one occasion and that was in northern Alberta in a saturated, poor peat bog landscape. The pitcher-plant obtains most of its nutrients through capturing of prey. However, prey capture is apparently inefficient, with only one percent of the visiting insects becoming trapped within the pitcher. Prey fall into the pitcher and drown in the rainwater that collects at the bottom of each leaf. Prey items such as flies, ants, spiders, and even moths, are then digested by an invertebrate community. Bacteria shred and mineralize available prey, making nutrients available to the plant.

Common bladderwort, another insectivore (also a carnivore), is an aquatic plant always found in the still waters of marshes, shallow lakes, and slow-moving watercourses that can be up to two metres deep. Below the water surface, there is an elaborate network of fine, leaf-like stems with inflated bladders. The trap door of the bladder operates at a speed ranked among the fastest plant movements known. Triggered by protruding hairs on the door, trap bladders open in a fraction of a millisecond, sucking the animal in and closing in about two-and-a-half milliseconds. Prey species include water fleas (*Daphnia*), nematodes, mosquito larvae, young tadpoles, and fish fry.

Sundews are another insectivorous plant, with round-leaved sundew being the most common. Two other species are oblong-leaved sundew and slender-leaved sundew. It is found in wet peat bogs throughout the Boreal Region of Canada and in bogs of the west coast. It has a circumboreal distribution, being found in most of northern Europe, much of Siberia, Korea, and Japan. The plant feeds on insects, which are attracted to the glistening drops of mucous, loaded with a sugary substance, that line the edge of its leaves. It has developed this adaption in response to its habitat, which is usually poor in nutrients or so acidic that nutrient availability is almost nil. The plant utilizes enzymes to dissolve the insects – which are stuck to the glandular tentacles – and extracts ammonia (from proteins) and other nutrients from their bodies. It had been assumed that insects were also attracted to the bright red colour of the common sundew. Still, studies using artificial traps have suggested that colour is not a factor in the attraction of prey.

SECTION 6

Working for Large Consulting Companies

In 2008 and 2009 when I was looking for a full-time job, there were few opportunities in BC. Ultimately, I found work in Calgary, and we moved there. Working for big consulting companies gave me a totally different perspective on the environmental consulting business. To stay employed, it was necessary to be chargeable most of the time. If I didn't bring in enough business or earn enough revenue, I would have been sent packing in short order. I became proficient at filing regulatory applications and receiving the necessary approvals required for all resource development and infrastructure projects.

MOVE TO ALBERTA

In 2008 I was defeated in my attempt at a fourth term as the Mayor of Esquimalt after serving as Mayor from 1990 to 1996 and again from 2005 to 2008. The morning after the election, a Sunday, Anne and I sat in bed talking about the future. We had lived in the same house in Esquimalt, BC for 27 years, and both agreed that it was time for a change. It was always a struggle working independently to find enough consulting work to pay all the bills. I decided to test the job market for vegetation ecologists.

The first thing I had to do was build a decent resume, one that would at least attract some interest from prospective employers. I hadn't produced a personal resume since the days when I graduated from BCIT. Anne's brother, Mark, was an invaluable resource, and between the two of us, I came up with a version that highlighted my vast years of experience and extensive skill set. In preparing the resume, I did a rough calculation of how many ecosystem plots I had completed in the 38 years of fieldwork and came up with the astonishing number of 10,000.

I spent hours scanning the internet for potential jobs and employers in BC, and the opportunities were slim. Picking out a few that I thought offered a chance for employment, I diligently fired off copies of my resume electronically. At the same time as doing this, we decided to put our house on the real estate market since we were convinced that a new job would likely require a move away from the Victoria region. December, January, and February came and went with no bites on the line for work. In the meantime, we received an offer on the house that was too good to refuse. So, there we were with the house sold and no job. Those were stressful times.

I am not sure how it came to be, but I had a phone conversation with Richard Pope of Dillon. During our talk I let him know of my plans to seek employment with an environmental company, and he asked me if I had considered Dillon. I told him no, since I had not seen any jobs listed on their website. He immediately asked me if I would be willing to come to Richmond for an interview. I jumped at the chance, and in a few days, I took the ferry from Victoria to Tsawwassen and met Richard and his partner Paul Schaap, another senior environmental scientist with Dillon. I spent two hours talking to them about my career, experience, and desire to have a full-time consulting job. They in turn asked me many questions about what kind of work I was seeking, what level of salary I thought I was worth, and if I would be willing to relocate. That was the key part of a possible job with Dillon. There was potential for a good job, but we would have to relocate to Calgary, Alberta.

I returned home full of optimism, and after talking it over with Anne, we decided that we should agree to an offer from Dillon (assuming it came). A week later, I had a call from Richard saying that Dillon would be delighted to offer me a job as a senior environmental scientist in the Calgary office. Shortly thereafter, I had a long discussion over the phone with their personnel manager, who was in Toronto, Ontario. It all happened quite quickly, but the job with its associated responsibilities and agreed-to-salary came to fruition. We were scheduled to be in the office in Calgary by March 31, 2009, and we frantically had to travel there before that date to find housing. Ultimately it all worked out, and on March 29, we loaded up our Subaru Forester with ourselves, travel luggage, and the three dogs (Milo, Lucy, and Chibby) and off we went to Alberta and a new life.

The Dillon office in Calgary included about 50 employees with a wide range of skills, including urban planning, project management, highway design, engineering, remediation, transportation planning, environmental assessment, and construction monitoring. I was excited to be starting work for a corporate entity since it was a completely new experience in my life after 38 years of working independently and for the government.

The first year at Dillon went by quickly. We settled into Calgary, exploring our neighbourhood, finding parks and green spaces where we could take the dogs, and getting to know the best shopping areas. At work, I was given small jobs to do, and I soon realized that the company structure didn't favour the workers at the lower levels. There were three partners at the top of the food chain, then a host of project managers who were mostly the senior staff of their respective disciplines. I was a simple peon beneath them, and for the most part, I was thrown the scraps of their projects, helping out where I could here and there.

One of the first things I did when arriving in Calgary was to apply for membership in the APB (Alberta Professional Biologists Association). Under the TILMA (Trade, Investment, and Labour Mobility Agreement), which came into being in 2009, a professional biologist registered in BC was entitled to automatic registration as a PBiol. (Professional Biologist) in Alberta upon application. British Columbia and Alberta mutually recognized and reconciled the rules that hindered the free movement of goods, services, and people between the two provinces. The agreement created a more open, competitive economy where goods, services, workers, and investments could move more freely between the two jurisdictions.

I did two interesting projects in 2009. The first was an EIA (Environmental Impact Assessment) to meet AT (Alberta Transportation) requirements for a new bridge to replace an existing one crossing the South Saskatchewan River on Highway 1 at Medicine Hat. During our field survey, we encountered prairie rattlesnakes. I have never enjoyed snakes much. Even the sight of a small one gives me the heebie-jeebies. It is a totally irrational response but apparently a common affliction. We were at the south end of the bridge when I almost stepped on a large snake that coiled up and started rattling. I must have jumped back about two metres as the snake quickly disappeared into shrubbery near the bridge abutment. Upon further examination, we observed three or four large rattlesnakes, known as the western rattlesnake or prairie rattlesnake, sunning themselves under the bridge. Ultimately, the snakes were removed and relocated before construction of the new bridge started[19] (see Footnotes and Appendix for details on the prairie rattlesnake).

The second project I did at Dillon in 2009 was a biophysical assessment for the Clearwater Heritage Park located a few kilometres west of downtown Calgary on Highway 8. Kiley Gibson, a wildlife biologist, was my assistant for the summer fieldwork, and I spent time explaining to her the nuances of ecosystem plot work. The Elbow River, which ran right through the middle of the park with a wide floodplain, eventually arrived to join the Bow River at the eastern extent of Calgary. The Elbow River has seen many flood events, and the historic flooding of 2013 (the worst of all time in the region) must have decimated the parklands. When I recently looked at the image of the park on Google Earth, nothing remained of it except for a couple of minor recreational features.

Water levels in the Elbow River are subject to fluctuations in the spring, determined by the depth of the snowpack, rainfall, and the timing of the season change. The annual runoff was originally referred to as a freshet (from the French word freschete or frais meaning fresh). This runoff produced periodic spring floods and regularly reduced water quality from the end of May to early July every year. Significant floods have occurred starting in 1883[20] (see Footnotes and Appendix for a chronology of major flood events).

In 2009 there was no spring flood on the Elbow, and it didn't take Kiley and me long to collect enough field data to allow for a comprehensive description of the park lands. Interestingly, portions of the floodplain (gravel bars) were characterized by the presence of calcareous pebbles and cobbles. Calcareousness can be tested for by applying a few drops of a weak HCl (hydrochloric acid). The acid causes a fizzing reaction with the $CaCO_2$ (calcium carbonate) in the limestone, which over time had been transported from the Rocky Mountains. Later, in the winter, we installed trail cameras at a few locations and obtained some good records of mule deer, coyote, and cougar using the habitat. Ultimately, the park was decommissioned and now sits vacant and unused by the city.

34 CASSIAR ASBESTOS AND JADE MINING

In the summer of 2010, Richard called me and said he had a biophysical project for us to do in northwestern BC near Cassiar. Mark Fraker, Kiley, and Laura Dilley, a young Dillon ecologist, came along to assist with fieldwork. I also wanted to teach Kiley and Laura more about plant identification and soils work. Kiley and I flew from Calgary to Whitehorse in the Yukon where we picked up a truck and drove south to Cassiar. The town had a unique history in the resource extraction domain of BC.

The discovery of asbestos in the area in 1950 led to the founding of the Cassiar Asbestos Company in the following year. The town was established in 1952, the same year the mining operation began[21] (see Footnotes and Appendix for more information). Mining in the area endured until 1992, when the mine closed, spelling the demise of the town. Today, the old apartment blocks at the east end of town are still used by mining exploration companies conducting underground gold mining at Table Mountain (formerly Erickson Gold) and base metal exploration in the immediate area. In the valley bottom was a massive slag heap of asbestos processing remains. The pile of material was at least 30 metres high and a few hundred metres long. I swore that at night the artificial hill glowed with a faint green colour.

There is also ongoing jade mining in the mountains around Cassiar. Jade mining is a lucrative business with strong demand for BC jade from China, Taiwan, and India. There have even been reality TV shows filmed about the jade mining business. Mark, Kiley, and I were curious about the jade. We had stopped in at the small Cassiar Mountain Jade Store at the intersection of Highway 37 and the Cassiar road to check out the jade items for sale. The raw jade is shipped to China, processed into saleable commodities, then returned to BC for sale in the store. So, one evening after work, we ventured down to an area just east of the townsite where we had seen massive boulders being loaded onto a flatbed truck. When we got there, no one was around, and there were no fences, so we just snooped a little.

There were huge boulders scattered about the work area, and most of them appeared dirty and nondescript. A few had been cut through with a large saw and we could see the dark green colour; jade doesn't show its real magnificence until it has been polished to reveal its vibrant green hue. The greens range from pale lime to moss and dark forest colour with every variation in between. There was also a pile of small cut pieces laying on palettes with a variety of patterns, including swirls, lines, and jagged lightning bolts. We decided we would come back the next day a little earlier to see if we could purchase a few pieces. The cut and polished jade products have an alluring beauty, sometimes with a translucent dreaminess and other times conjuring up primal images of ancient plant material[22] (see Footnotes and Appendix for details on jade mining). But we had a job to do in Cassiar.

We accessed the mostly mountainous area primarily by helicopter, flying up to more than 2,000 metres in elevation. The vistas were spectacular, highlighted by steep alpine slopes, fellfields and ice fields, subalpine meadows sprayed with colour, and fields of white and pink mountain-heather. We knew there could be bears about, so whenever we departed from the chopper, our bear spray canisters were always attached to our hips.

One day when we were in a low-elevation fen, with water at the surface and a perimeter dominated by willow and brilliant red-osier dogwood, we startled a cow moose and her twin calves. We were in the middle of the wetland, and she was about 30 metres away, munching on aquatic herbs. Momma was immediately on high alert. Her ears went flat on her head, and she let out a loud bellow as she shook her massive head with its bulbous nose. We were in a vulnerable position, but luckily for us, she spooked. With a loud snort and another head shake, she jumped from the water at the shoreline into the young white spruce forest. We had a good look at the young calves before they followed the cow. They were about one and a half metres tall with spindly legs and reddish-brown with darkening on their hocks. They both bleated at the cow. She gave an answer, and in an instant they followed her into the forest.

Moose are one of the most amazing creatures to see up close. I have seen many cows and bulls over my career, but this was the closest I had ever come in person to them. Moose are found right across Canada in forested environments, although I have even seen them in the fields on the prairies in Saskatchewan. Big bulls can weigh more than 700 kilograms, and the giant Alaskan moose can have a rack more than two metres across.

Later that year, Dillon sent me to participate in a golf tournament in Leduc, south of Edmonton. My power cart partner worked for EBA (Elmer Brooker and Associates, which was bought by Tetra Tech Inc. in early 2011), and he was interested in what I did for a living. When I told him that I was a professional biologist specializing in vegetation ecology, he asked me if I would consider moving to Edmonton. He talked to the senior environmental scientist at his office by phone, and during the round of golf, he told me that she would be interested in meeting to talk about hiring me. It was a serendipitous occurrence that the only golfer from EBA would end up riding in my cart.

When I got home, Anne and I talked about the possibility of moving again. She wasn't enjoying our time in Calgary, a busy, fast city more suited to young people, and she had an aunt and cousin living in St. Albert, just northwest of Edmonton. So, the next day I phoned Kelly Ostermann, EBA's senior environmental scientist, and we arranged for an interview in a week. After meeting with her and talking about a potential job, I waited to see if an offer would follow. Sure enough, within a few weeks, EBA offered me a position as Senior Biologist in their Edmonton office with a salary increase of $18,000 per year. I took the job, and in March 2011, we moved to St. Albert.

35 THE HIGHWAY OF DEATH

EBA was primarily an engineering company with an environmental component. Our group included professional agrologist and group leader Kelly, aquatic biologists Eddie Noton and Michael Villimek, soil scientist Mark Fawcett, environmental scientists Amy McLaughlin and Tara Bodeux, and junior biologist Teia Clement (no relation to me). Part of the reason for Kelly bringing me on board was to train and mentor the junior biologists.

My first project was to do all the environmental assessment work required for a large overpass (41st Avenue SW) on the Queen Elizabeth II highway just south of Edmonton. The highway runs between Edmonton and Calgary for almost 300 kilometres). I was responsible for collecting data and mapping of wetlands within the project footprint, as well as completing weed and tree surveys, rare plant surveys, and soil descriptions. Once all the data had been collected, EBA was required to complete and file a CEA (Canadian Environmental Assessment) report as directed under the *Canadian Environmental Assessment Act* of 1995; the act was revised in 2012 and again in 2018. The act ensured that any large-scale infrastructure or resource extraction projects completed their due diligence regarding all environmental components, including vegetation (including wetlands and rare plants), soils, wildlife (with a focus on listed [rare or endangered] species), fish, and water.

I enjoyed the challenge of writing the assessment report. For the most part, there were existing templates and prescriptions which dictated the format and contents of these reports. Along with the report, EBA also needed to file regulatory applications with federal agencies (to be compliant with the *Fisheries Act, Navigable Waters Act,* and *Species at Risk Act*) and provincial agencies (the *Water Act, Public Lands Act, Soil Conservation Act, Weed Control Act, Wildlife Act,* and *Environmental Protection and Enhancement Act*). In addition to those acts any construction must consider the potential for impacts to nesting birds during the breeding bird season (according to the federal *Migratory Birds Convention Act, 1994*), which runs from March 31 to August 15. It was a steep learning curve, but I enjoyed the challenge of acquiring a completely different skillset from the ones I had been immersed in over the first 40 years of my career.

Eventually, the reports and applications were all completed and submitted, and then we waited for an interminable amount of time as the regulators processed all the necessary paperwork. It took almost two years for the entire construction schedule to be completed. Now when I drive under the massive overpass on my way to Calgary, I feel a sense of pride and accomplishment at my involvement with the EBA staff and our part in the successful construction of the major interchange

From 2011 to 2013, I was also responsible for carrying out the construction monitoring for another interchange project on Highway 63 north of Fort McMurray. It meant I had to drive up Highway 63 once a month and stay in Fort McMurray. Both the driving and staying in downtown Fort McMurray were unpleasant experiences. Highway 63 was known as the 'Highway of Death.'

With the advent of the oil sands boom in the 1990s and early 2000s, the highway saw increasing traffic, with a significant number of commuting workers using the route. Between 2001 and 2005, there were more than 1,000 collisions on the highway resulting in 25 fatalities and 257 injuries. Regular users of the highway suggested that bad driving was to blame for most of the collisions, with impatient drivers making poor decisions. Shift work is typical in the oil sands business, and residents of Fort McMurray expressed the opinion that such vehicle operators are more likely to be travelling at excessive speeds or under the influence of alcohol.

I personally experienced trucks passing me at 90 kilometres an hour in construction zones that were posted at 50 kilometres an hour. By late 2011, only 19 kilometres of the twinning project had been completed. On April 27, 2012, a horrific head-on collision on the highway near Wandering River killed seven people, and there was a huge public outcry. The Alberta government responded immediately, announcing that the rest of the twinning project would be fast-tracked.

Later that year, we received word that our transportation division had been successful in bidding on projects involving AT and the gargantuan task of twinning of the remainder of Highway 63. The total length of the highway is 434 kilometres but the section being twinned ran 285 kilometres from near Lac La Biche to Fort MacMurray. We were responsible for two sections of the highway, one 200 kilometres south of Fort McMurray and just north of Wandering River, and the other around the Hangingstone River which was closer to Fort McMurray. As with other AT projects, we were responsible for obtaining the necessary regulatory approvals for construction to proceed. We did all the associated fieldwork, filed the applications, and received the approvals in short order. Construction proceeded full steam ahead.

Similarly, as with other projects, I was responsible for the monthly construction monitoring. During these trips, I had to ensure that the construction companies were in full environmental compliance. Any transgressions, such as fuel spills or sediment releases into watercourses, could result in a shutdown of work issued by federal or provincial regulators.

In the spring of 2013, I had driven the highway to Fort McMurray to carry out my regular monthly check on the Highway

686 interchange. It was an unseasonably warm day for the time of year. There was still snow on the ground, which was melting at an accelerated rate. On my way through Wandering River, at about 11 am, I could see streams running down the sloping fields above the highway towards the construction area. I was concerned about the possible occurrence of sediment loading into the river. Sure enough, when I made the trip back, late in the afternoon, there were torrents of water rushing off the slopes and crossing the highway on both sides of the Wandering River bridge. The highway was flooded, and sediment material was pouring into the river. I noticed construction workers frantically trying to divert the water, but it was no use. Ultimately the Wandering River was flooded with huge volumes of soil material carried in the freshets, and we were obliged to file a non-compliance notice with Alberta Environment and Sustainable Resources Development. The ministry was outraged and threatened to shut down all work, but somehow, we were able to remedy all the erosion and sediment control measures and carry on with the work. All the twinning work on Highway 63 was completed by 2016 at a cost of more than $2 billion dollars. Since the twinning, vehicle collisions have declined significantly.

In October of 2013, I was working in my office, and out of the blue, I received an interesting email from a 'headhunter.' The message simply asked, "I have a client who is interested in your skillset, would you be willing to consider talking to them?" I thought about it for a few minutes; I was happy at EBA-Tetra Tech and had no reason to look for another job. I decided there was nothing to lose and everything to be gained and wrote back, "Maybe." Shortly, another message came back, "What would it take for you to consider coming to Stantec?"

Stantec was a huge engineering and environmental company with offices all over North America, one of EBA-Tetra Tech's main competitors. My current annual salary was in the order of $120,000 gross. I thought about it again, then wrote "$140,000 per year." He didn't respond, so I assumed that was the end of our electronic conversation. After an hour, he messaged again. "Okay, my client is interested and will contact you for a potential interview date and time." I still didn't know if Stantec was serious, but within a week, I received an invite to an interview at Stantec's downtown Edmonton office. I was very clear with the interviewers that I would only be willing to change companies if they could commit to allowing me fieldwork opportunities. I was not prepared to move into an administrative or project management role. The interview lasted about an hour, and when I left, I had a positive feeling, but I still had no idea how serious they were. Ultimately, they made a generous job offer which included an annual salary of $150,000. Everyone at EBA-Tetra Tech was sad to see me go, and some of them grumbled that I was going over to the 'Dark Side,' but in November 2013, I started working in the Stantec office.

PRAIRIE GRASSLANDS AND WOOD TICKS

Stantec started as a one-man operation in 1954 run by Don Stanley in Edmonton when it was called D.R Stanley and Associates. When I joined them in 2013, the company had morphed into a massive, publicly-traded corporation with around 17,000 employees. In 2014 they reported gross revenues of $2.5 billion.

We had a good vegetation ecology crew at Stantec, including my supervisor Jessica Feschuk, junior environmental scientists Nadine Clifton, Emony Nichols, Jennifer Bles (Jenn later became Jennifer Gosse), and intermediate ecologist Jenn Lange. Later in 2014, senior vegetation ecologist Dave Downing joined our group on a part-time basis. Just a few years younger than me, Dave was the author of numerous scientific reports, including a series of comprehensive biophysical manuals for the Northwest Territories and a publication entitled Natural Regions and Subregions of Alberta. I enjoyed his cynical and caustic wit. The soils group that we often worked with included Ivan Whitson, Kaylie Forster, Christiane Brouwer, and Brett Leoppky. The Stantec environmental group included about 120 employees.

In early summer, Nadine and I were sent to work on the Uplands Pipeline project (a Trans Canada project that never proceeded) around Carlyle, Saskatchewan, close to the Manitoba border. It was the first time that I had worked in a 'true' prairie grassland environment, and it was challenging and stimulating. Our job was to conduct rare plant surveys on sections of the proposed pipeline route. For my entire career, I had collected vegetation and soil data on hard copy data forms, and this was the first project where we were obliged to use electronic data loggers. I was not impressed with their performance. The screen was small, only about 10 x 13 centimetres, and the units would often freeze up or not accept data. The worst part was in bright sun, it was almost impossible to see the details of the loaded imagery.

I had always imagined that working on the prairies would be easy. How wrong I was. I was used to doing field surveys in mountain and coastal environments, with steep slopes, never-ending shrub thickets, and the constant risk of bear encounters. Here on the prairies, we could see in all directions for 360 degrees. There was a reason it was known as 'Big Sky Country.' We literally saw thunderstorms coming from more than 20 kilometres away. There was no danger of running into bruins, but other threats had to be taken seriously. Barbed wire fences had to be negotiated, and abandoned tangles of wire were often hidden in long grass or just lying on the ground. Gopher holes were everywhere, and it would be easy to sprain an ankle if one stepped into these little traps. On more than one occasion, we had to cross a farmer's field where cows, and an occasional truculent bull, were giving us the evil eye. In the grasslands, there was virtually no shade and nowhere to seek respite from the incessant heat and intense sun. Lack of drinking water further exacerbated heat-related stress.

I spent a good amount of time just becoming familiar with the prairie flora and learning many new species. Unlike in Alberta and BC, there were no published field guidebooks for plant identification. There is an archaic flora manual called Budd's Flora of the Canadian Prairie Province (Budd 1987) which we used, but it had no illustrations to help confirm identifications.

It was late spring, and we were in wood tick country. The wood tick (*Dermicentor variabilis*) does not normally carry Lyme disease, although it does infect people with other disease-causing pathogens, one of which can cause Rocky Mountain spotted fever. Although many people become ill within the first week after infection, signs and symptoms often don't appear for up to 14 days. Initial signs and symptoms of the disease often are nonspecific and can mimic those of other illnesses. Symptoms include high fever, chills, severe headache, muscle aches, nausea and vomiting, and confusion or other neurological changes. Lyme disease, a debilitating illness that is notoriously difficult to diagnose, is spread by the deer tick (*Ixodes scapularis*) common in the eastern part of North America. Wood ticks are about three to five millimetres long with a dark (almost black) body, while deer ticks are considerably smaller with an orange abdomen. Wood ticks occur in the dry interior of BC, and the prairie environments of Alberta, Saskatchewan, and Manitoba. With climate change occurring, wood ticks have been reported as far north as St. Albert, in Alberta. Ticks secrete small amounts of saliva with anesthetic properties so that the animal or person cannot feel that the tick has attached itself. Because of that, unless one feels a tick crawling on their skin, finding the tick can be difficult. If the tick is in a sheltered spot (such as the groin, scalp, or armpit), it can go unnoticed and can slowly suck the blood for several days. We would do regular tick checks every hour and often find ticks on the outer layers of our clothing. One night in the hotel, I still do not know why; I awoke at around midnight to find a tick crawling up my stomach. I had a second trip to Carlyle about three weeks later with Jenn Lange, and she would nearly have a panic attack if I noticed a tick crawling up her clothing. She swore she would never come back to Saskatchewan.

Later in the summer, I was dispatched to BC to assist with a TEM survey for the Prince Rupert Gas Terminal project. The client was also Trans Canada. All our work was done by helicopter, with the pilot staying with us all day. Stantec was partnering with EDI (Environmental Dynamics Inc.), and I was paired with Rachelle Robitaille, a wildlife biologist. We flew the pipeline route, which was predominantly routed in mountainous terrain, and landed at polygons that had been pre-mapped, where we completed an ecosystem plot. We did part of the work out of Smithers, then moved to Terrace. We were complemented by two Nisga'a crew members for the work. James was assigned to help dig soil pits; the second person, Neil, was there as the Wildlife Monitor. He carried a rifle or shotgun, and his job was to watch for wildlife signs and protect us in the event of a bear (or other carnivores) encounter. The work went well, and we were progressing from east to west to ultimately end up near the marine environment on the Nass River. The Nass flows from the Coast Mountains for about 380 kilometres to enter a side inlet of Portland Canal. Near the end of its long journey from the mountains, the floodplain is expansive, almost two kilometres wide. There are alder forests, estuarine plant communities, skunk cabbage swamps, and Sitka spruce trees. Perfect habitat for grizzly bears. When we were working on the floodplain, we saw obvious and fresh grizzly signs almost everywhere we went. Neil, the wildlife monitor, did a good job of warning the bears of our presence. One morning, when the clouds were hanging low in the valley and the light was soft and grey, we were looking for a helicopter landing spot close to a potential sample site. As the pilot made a sharp turn over the estuary, I suddenly saw a huge brown shape, followed by two smaller ones, running through the tall sedges. I yelled into the intercom system, "Hey, we've got a huge grizzly sow with two clubs below us"! The pilot yelled into his mic, "Where"? I fired back at him, "About 10 o'clock." He quickly swung the machine around so that he could see the bears. Sure enough, there they were, heading for the cover of trees, which they reached in no time. Rachelle immediately said that we shouldn't land anywhere nearby, and I agreed. The EDI protocol when grizzlies were near sampling sites stated that if there was a high risk of having a bear encounter, then the sample area should be abandoned. Since we had no alternative area to survey, we headed back to Terrace. I think the crew realized how close we had come to putting ourselves in serious peril.

COASTAL GAS LINK FIASCO AND WIND TURBINES

In 2015 we were part of a huge team working on the CGL (Coastal Gas Link) project, another Trans Canada (now called TC Energy) pipeline. The pipeline was planned to move natural gas between Dawson Creek and Kitimat for a length of 670 kilometres. The section we were working on was centered south of Houston, BC. On the first day of our work, we were obliged to participate in an orientation session held in a huge gymnasium in Houston. The meeting, where there were more than 300 people in attendance, included presentations by Trans Canada, the construction company, and First Nations representatives. Our fieldwork, which was mostly done by truck using road access, consisted of locating ecosystem plots and collecting vegetation and soil data. To do that, though, we were required to take a 'team' of interested participants with us. Our team consisted of myself and one other field worker from Stantec, two First Nations technicians, two archaeologists from CH2M Hill, a wildlife monitor, and four elders from the affected bands. As we drove the roads to our sampling sites, our cavalcade of four trucks progressed slowly to ensure we didn't lose contact with any of our vehicles on the way. Walking through the bush with 11 people was even more laborious. Some of our team members were not in the best shape, with little or no bush experience, making progress challenging. The wildlife monitor carried a shotgun but had an irritating habit of using a bicycle horn which he honked incessantly to warn any bears of our impending approach. Needless to say, our quota of daily ecosystem plots was never achieved.

Most of the local bands supported the project, but the Wet'suwet'en hereditary chiefs were not on board. Beginning in 2010, the hereditary chiefs and their supporters set up barricades and checkpoints along the Morice West Forest Service Road that provides access to the construction of the CGL pipeline that threatened their territory. We were told by Trans Canada that under no circumstance were we to engage with any protestors if our progress was impeded. In other words, if our route was blocked, we were simply required to turn our trucks around and leave. During our field session, we never encountered any protestors or had our passage stopped. A detailed chronology of events leading up to the massive national protests (in 2020), in support of the Wet'suwet'en hereditary chiefs, is provided in the Appendix[23].

Later that summer, I was sent to the Stantec office in Regina to pick up my field assistant, Amber Smith, and project materials for a rare plant survey of a proposed wind power project in southwest Saskatchewan. The interest in alternative renewable energy sources, such as wind and solar, is a growth industry and one that will only expand over time. The project we were working on involved the potential erection of more than 70 wind turbines on a hummocky landscape about 90 kilometres west of Moosejaw. Our job was to conduct rare plant surveys on the large mounds where the turbines would be erected.

The terms 'wind energy' and 'wind power' both describe the process by which wind is used to generate mechanical power or electricity. This mechanical power can be used for specific tasks (such as grinding grain or pumping water), or a generator can convert this mechanical power into electricity. A wind turbine turns wind energy into electricity using the aerodynamic force from the rotor blades, which work like an airplane wing or helicopter rotor blade.

Typically, the landscape of the southern prairies is flat or rolling, but the area where the wind power project was planned was markedly different. Due to meltwater deposition during deglaciation, huge mounds approximately 300 metres in diameter and 50 metres tall occurred over an expansive area. At each mound, where the turbines would be erected on top, we were required to walk in concentric circles, spaced about 10 metres apart, covering an area about 100 metres across.

The weather during our survey work was excessively hot, with daytime temperatures exceeding 37 degrees Celsius. There were no trees to afford shade, and we had to take serious protective measures and implement safe work strategies to ensure that we didn't become seriously sunburnt or dehydrated. To add to our misery, smoke from forest fires in BC clouded the sky every day. We would leave Moosejaw around 5 am and try to finish all our work before 2 pm when the sun was at its most intense. I wore a straw hat to shade my head, a light shirt and cotton pants, and a cloth soaked in water around my neck to beat the heat. Every two hours, we would go back to the truck, pump up the air conditioner, and rest for about 15 minutes before resuming work. We never found any rare plants, but by the end of our field trip, we were experts at the local flora.

On the way to our project area, from Moosejaw, we passed through the historic town of Chaplin, which is home to a large salt (sodium sulphate) mining operation. Construction started on a salt processing plant (owned by Saskatchewan Minerals Inc. at Chaplin 1947. As the demand for salt cake grew, so did the mining operation. In 1983 the company made a major investment by installing state-of-the-art multiple-effect evaporation equipment at the Chaplin operation. The equipment not only allowed for them to produce high-grade sodium sulphate but to do so in a very efficient manner. Saskatchewan has 21 major sodium sulphate deposits with five operational salt mining plants. The product is primarily used for detergents (laundry and dishwashing soap), carpet deodorizers, corn starch, glass, dyes, textiles, tanning, and in the chemical industry.

Chaplin Lake, which is close to Chaplin, is home to thousands of shorebirds. Over thirty species have been documented there, including endangered burrowing owls, ferruginous hawks, and long-billed curlews.

In the early fall, I was once more summoned to Saskatchewan to carry out wetland assessments for the Uplands Pipeline project. Amazingly, mosquitoes were quite common on the prairies, and we were subject to clouds of them. At wetland locations along the pipeline route, we were required to map the wetland extent (my assistant Mark would do this with a GPS unit) and compile a complete list of plant species present. We did find a rare member of the Sunflower family, common beggar-ticks (*Bidens frondosa*), with some regularity. It is ranked as rare by the Saskatchewan Conservation Data Centre. On our last day of work, we were wading on the edge of a wetland when I heard a strange noise. It was a kind of moaning sound, which I first interpreted to be a toad or frog. I said to Mark, "Did you hear that"? He said, "No, what"? I told him to be quiet and listen. Sure enough, there it was again. A weird low moan. We left the wetland and climbed up over a small ridge, and there in front of us was a poor cow that had walked into a wetland and gotten horribly stuck. Mark and I felt so sorry for the poor beast that we were determined to try and save her. It took us a while, but we were able to finally track down the farmer and call him on my cell phone. He thanked us and said he would be there as soon as possible. We waited about 30 minutes until he showed up. He told us that the cow had been missing for three days, which meant she was probably severely dehydrated. We could see massive lumps all over her back where flies had laid eggs in her hide. Within a few minutes, he had extricated her with his tractor, but she wasn't able to stand up, and we left. I don't know whether she survived her ordeal, but at least we had done a good deed and helped relieve her untenable situation.

In the summer of 2015, Stantec started jettisoning staff as if they had the plague. People were being let go on a regular basis. By the end of the year, most of the vegetation ecology and soil groups were gone. Jessica and I were the only ecologists left. Morale was terrible in the office, which now included only about 60 people. I figured it was only a matter of time before my number came up.

SECTION 7

Working Independently for the Third Time

With Stantec wanting to make a casual worker, it was an easy choice to become an independent environmental scientist again. At the age of 67, the opportunities for good-paying consulting jobs were limited. Anne and I were receiving federal pensions now, so I didn't really need the income of a full-time job. However, the consulting contracts were few and far between, so I ended up doing other jobs with a drilling company and Canadian Tire.

ALONE FOR THE THIRD TIME AND FIELDWORK IN THE ARCTIC

Sure enough, in March 2016, Stantec decided to change my employment status. I had just come back from five days of golfing in Palm Springs. My boss called me to a meeting, and I knew when I saw the HR staff person in the room that something was up. I was informed that they were going to put me on casual status, which meant I would have to wait until there was work before they needed my services. There was no attempt to offer me a reduced workweek or lower salary, just a decision to make me casual. In the consulting business, it's all about chargeability. If you are not bringing in enough work, then you are deemed to be expendable. The amazing thing was in 2015, I was 75% chargeable. At my hourly chargeout rate of $246 an hour, that meant that I had generated approximately $365,000 of gross revenue for the company in 2015. Deducting my salary of $150,000 meant that the remaining $215,000 dollars was profit and money to cover carrying costs. Not bad for one year. It still wasn't good enough to ensure my ongoing employment with Stantec. I politely declined the offer of being a casual worker, and once again, I was alone.

Don't misunderstand me; I enjoyed my two and a half years with Stantec. True to their word, I was afforded the opportunity to be an active field ecologist for a wealth of interesting and challenging projects. I had worked in three provinces with an array of different and stellar resource scientists. The reality was that working for Stantec was like being part of "the collective," the term used for describing the Borg in Star Trek or a worker bee in a huge hive. The administrative requirements, although a necessary evil, were odious and overbearing. Meetings were constant and boring. I did not miss working for the corporation, although I must admit I missed getting the large, monthly paycheques.

I formed my own company, Chris Clement Eco Solutions Ltd., and was ready to look for small consulting projects, which brought me to the beginning of this book and the carbon sequestration project on Quadra Island in 2016. I took most of the year off, drawing on the retirement funds I had accrued from EBA and Stantec.

Cambridge Bay field crew; from left – Robert Fraser, Johann Wagner, Carla Schmitt, Ian Olthoff, Chris Clement, Angulalik Pedersen, Aila Pedersen and Christopher Bolduc

However, I was engaged for one very cool project by Stantec in the summer to carry out research on shrubification, which is the northward advance of shrub species in the Arctic as a response to climate change. We had been awarded the project in late 2015, and Stantec needed me to do the fieldwork in 2016. The concept of studying the effects of climate change on vegetation in the Arctic was conceived by Dave Downing, but he left Stantec before we could articulate a meaningful proposal.

The proposal, submitted in response to a Request for

Submissions from PKC (Polar Knowledge Canada), included the establishment of permanent research plots near Cambridge Bay, Nunavat. Cambridge Bay is 1,788 kilometres north of Edmonton, located on the south coast of Victoria Island, just across from the Nunavut mainland. In 2016 the population was 1,766.

On July 26, I boarded an old Boeing 737 in Edmonton and flew to Yellowknife, Northwest Territories, where we had a short stop before continuing to Cambridge Bay. The flight was ridiculously expensive, over $2,500. Interestingly the plane only had room for about 20 passengers as it had been converted into a cargo carrier, with a wall between us and the cargo hold. When I arrived at the Cambridge Bay Airport, there was no one there to meet me. I waited for a while. Waited more time, and still no pickup, so I called for a cab. After another long wait, I threw all my gear in the dilapidated taxi that finally showed up and headed for the PKC office in town. When I arrived, there was no one about. I knew that this far from civilization, time sort of slowed down. Southerners called it 'Northern Time.' With not even 2,000 people living in town, no traffic lights, an absence of paved roads, and never-ending sunshine, what would be the point of rushing around or getting stressed out? I was immediately struck at how barren the landscape appeared. This was as far north as I had ever been, and I had never even made it as far as Yellowknife. There were no trees, only a few short, weather-beaten shrubs, and endless hummocky, Arctic tundra. In late July, the sun set at around 11:16 pm and rose at 2:57 am, which meant that it never really got dark.

Eventually, a wiry, energetic fellow showed up, and I introduced myself. He was Dr. Johann Wagner, the Senior Ecologist of PKC who also worked at the CHARS (Canadian High Arctic Research Station) every summer. He took me into the office where I told him what I was in Cambridge Bay to do. He had forgotten that I was coming, but once I expounded on the proposal and the planned fieldwork, he immediately offered to help with the data collection. He then introduced me to the rest of his team, including environmental field technician Angulalik Pedersen, environmental assistant Aili Pedersen, and summer student Christopher Bolduc. I told Johann that I wanted to start work the next morning, to which he agreed. After chatting for some time and looking at the sampling frame, which he pulled out of a dusty closet, he told me he would drive me over to the CHARS residences. The building was brand spanking new and part of the new research station complex that was to be christened by Prime Minister Trudeau a week after my departure date.

I was not sure when the rest of our crew, the staff from CCRS (Canadian Centre for Remote Sensing), were arriving, so I settled into the apartment. It was like a spacious condo, which had never been lived in, with four bedrooms, a huge living room, and compact kitchen. Once I had stowed my gear, I decided to go for a walk to check out Cambridge Bay. I had already scoped out the general lay of the landscape on my trip from the airport into town, but now I was itching to get out and look at the flora up close.

I walked into town and immediately noticed that most of the residences were built on pilings. I am sure this was to deal with the issue of melting permafrost. With climate change, the issue of melting permafrost is causing problems right across northern Canada. Buildings are subsiding and tilting, slopes are sliding or failing, erosion and cratering is rampant, and greenhouse gas emissions are skyrocketing. It seemed every residence had a dog or two (mostly Huskies) tethered in front of it, and an ATV (all-terrain vehicle, also known as a 'quad') and snowmobile (also known as a 'sled') parked beside it. Many of the older houses were askance, bent this and that way, worn by years of battering from the snow and wind. There was no protection from the elements. The community appeared to have been dropped down on the tundra from the sky. There were functional buildings, rough dirt and gravel roads, no sidewalks, and no landscaping. The only landscaping was provided by the natural environment. Once I stepped outside the limits of the community, I was on the Arctic tundra. Quads were roaring up and down the streets, causing constant clouds of dust.

Coming from a life of working in mostly treed environments, being exposed to the endless tundra was stunning to me. Even on the prairies in the grasslands, there was a diversity of ecosystem and landscape types, including shrubby and treed gullies or draws, a variety of different wetlands, sprawling floodplains, and thick aspen forests. Here, there was just the tundra that was characterized by miniature ecosystems all crammed into a height of 30 centimetres. I could imagine the wind howling across the landscape during winter, daring any plants to raise their tiny heads. The wind seemed to be constant, even during my time there. To really appreciate the tundra, you had to get up close and personal with it. When I got down on my hands and knees, I could suddenly comprehend the amazing diversity that presented itself. Most of the vascular plants (those containing phloem and xylem which are conductive tissues) were short and compact; they were complemented by a host of lichens and mosses which hugged the soil and stones. The ground was hummocky, with the little mounds being a maximum of 20 centimetres high and 12 centimetres across. The hummocks are caused by the constant process of frost-heaving (called cryoturbation) that goes on for more than eight months of the year. I laid on the ground and peered across the hummocks. The true character of the tundra revealed itself to me. I didn't know the names of many of the plants I was looking at but over the next two days of working on the research plots I became familiar with them.

I wandered into the local grocery store and bought a few provisions at a cost two or three times the price I would have paid back home, then headed back to the condo. By the time I got there, the CCRS crew had arrived, and we exchanged introductions and pleasantries. Research scientist Dr. Robert Fraser was the remote sensing team lead; he was joined by remote sensing scientist Dr. Ian Olthoff and physical scientist Carla Schmidt. They were going to be using sophisticated drones (the technical term is UAVs [Unmanned Aerial Vehicles]), with cameras mounted on them, to shoot images of the research plots. The images were going to be analyzed to determine how accurately they could predict cover values of the plant species groups.

The next morning Johann arrived around 8 am to pick us up, and we headed down the road to Greiner Lake. Angulalik, Aili, and Christopher followed us in another PKC truck. We ended up locating our two research plots about 10 kilometres northeast of Cambridge Bay. A detailed account of the rationale for the research and the methods of doing the work, taken directly from the technical report that Robert and I wrote and submitted to PKC, is provided in the Appendix[24].

We spent July 27 laying out the first research plot and collecting vegetation data. It was too windy for the UAVs to be flying so we delayed that part of the work until the next day. The next morning dawned calm and warm (perfect for black flies and mosquitoes), and we headed back down the rough track past the first plot that lacked any significant shrub cover. We drove for another kilometre and found a good sample site with patches of willow. The PKC environmental staff helped me lay out the plot, and we proceeded to document the plant species. The process of using the metre frame and wire to record intercepting species was time-consuming but reliable. Between the two plots, I documented three species of willow, as well as alpine bistort, northern white mountain-avens, and Inuit carrot. The Inuit carrot (*Oxytropis maydelliana* [a member of the pea family]) has a long history of traditional use[25] (see Footnotes and Appendix for details on the use of the Inuit carrot). The other species that were common included large-flowered lousewort, purple saxifrage, polar grass, and seven sedge species. The lichens and mosses included pale Iceland lichen, dark Iceland lichen, orange rock lichen, reindeer lichens (species were not determined), whiteworm lichen, scorpion moss, and two kinds of crust lichens (they grow directly on the soil).

At the end of the day, Robert, Ian, Carla, and I went out for dinner and had a fine feast of fresh Arctic char. Cambridge Bay has a commercial Arctic char processing plant. The char are an important component of the social connection, way of life, and food requirements of the Inuit people. Cambridge Bay is known as Ikaluktutiak, which in Inuinnaqtun translates to 'Good Fishing Place' and reflects the enduring historical and cultural connection the people have with char. In 2012 the operation employed 28 residents. The char fishery has a huge beneficial impact on the community and Nunavut. In 2012, the commercial harvest exceeded 48,000 kilograms, and the market value for all forms of char produced was estimated at more than a million dollars.

The next day I had to return to civilization, and I was sorry to say goodbye to the CCRS and PKC staff. The survey work on the Arctic tundra had been an eye-opening experience, one that I thoroughly enjoyed. My only regret was that I never saw any of the truly Arctic creatures, such as barren-ground caribou, muskox, or Arctic fox. I went back to Cambridge Bay the next year to repeat the data collection process, and our proposal did include additional data collection after a five-year period. To date, I have no idea if the next scheduled research will be implemented.

I did one small job for Stantec after the Arctic excursion that related to the catastrophic fire that engulfed Fort McMurray. On May 1, a wildfire began southwest of Fort McMurray, Alberta[26] (see Footnotes). On May 3, it swept through the community, forcing the largest wildfire evacuation in Alberta's history, with upwards of 88,000 people forced from their homes. Rampaging through Fort McMurray, like banshees from hell, the conflagration destroyed more than 2,000 homes and buildings. The fire continued to spread across northern Alberta and into Saskatchewan, consuming forested areas and impacting Athabasca oil sands operations. One entire work camp was obliterated. With an estimated damage of $9.9 billion, it was the costliest disaster in Canadian history. The fire blackened approximately 590,000 hectares before it was declared to be under control on July 5, 2016. No official cause was ever determined.

I went to Fort McMurray with Ben Steele, an environmental technician employed by Stantec, to carry out some vegetation surveys on a portion of the burnt area near Saline Creek. It was weird being in Fort McMurray, which was like a ghost town. There was hardly any traffic on the roads, few people were about, and we could see the burnt neighbourhoods with the charred ruins of houses. Because of the high amount of soot and dust still in the air, we were required to wear hazmat suits and respirator masks during our work. With the heat and haze, it made for uncomfortable working conditions, and we would sweat profusely. Wearing the suits made me feel like I was in a science fiction movie, working in a post-apocalyptic landscape. The grim reality set in any time I caught a glimpse of the remains of a former neighbourhood, and the houses were a skeletal reminder of the families who had lost everything. It was amazing to see that within two months of the area being burnt, there was already a profusion of plant species such as prickly rose, blue joint, false sarsaparilla, and bunchberry popping up from the charred surface layer. The fire had gone through the forest so fast that it effectively killed the trees but only burnt the top of the duff layer (moss and humus on top of the soil surface). The plant roots that were deeper had survived. After a few days of suffering from the heat and dust, I was glad to leave Fort McMurray behind.

30 DIRECT HORIZONTAL DRILLING AND THE GARDEN CENTRE

In spring-2017, my consulting job prospects looked grim. A golfing buddy, Roger Running, asked me if I would be interested in working for DHD (Direct Horizontal Drilling). His brother-in-law, Lon, was a part-owner of DHD, a company that specialized in drilling under watercourses for pipeline projects. I thought, why not? I had nothing to lose, so I submitted a resume. Lon was impressed by my experience and hired me on the spot. The job was being a crew driver for the company, which involved transporting the drilling crews back and forth between hotel, or camp, and the drilling rig. The drilling rigs, worth over a million dollars, were set up on both sides of the river or creek where a pipeline needed to cross beneath. Using a sophisticated computer system and a massive drill bit, they sometimes took months to complete the drilling. The crew consisted of five men: a driller, a machine operator, a mud man, and two rig hands. They are directed by the rig manager, who lived in a trailer at the drilling site for the duration of the work. The labourers were also known as roughnecks or 'rig pigs,' and they were most often very crude, typically young and immature, foul-mouthed individuals. The rigs ran round the clock, with two crews working 12-hour shifts at each rig site. I diligently obtained all the necessary work tickets and then waited through the summer for a call from the company. I was a fill-in driver and would be spelling off regular drivers when they needed a break.

It was early April, and I needed some form of work to help pay the bills. I decided that a good opportunity might be to work for the Garden Centre at the Canadian Tire in St. Albert. One of the problems with applying for jobs was that most outfits only accepted electronic submission of resumes. Once they figured out how old you were, they threw your info into the round file (in the wastebasket) or deleted your information. I knew that Canadian Tire still accepted hard copy resumes with job applications, so I went to the store and got an application form. The day I submitted my form, with a specific request for work in the Garden Centre, they called me. I guess whoever had looked at my resume was blown away by my list of accomplishments and qualifications (professional biologist, vegetation ecologist with over 45 years of experience, plant expert, and professional forester). They asked me to come down for an interview which I did the next day and they hired me on the spot. My boss, a jovial, matronly lady with a great sense of humor, was Tammy Morin. We still had snow on the ground, so she told me to report for work on April 23, when we would start getting the Garden Center in shape to receive the upcoming plant material.

When I showed up, we had to clear out the greenhouse of all extra stock that had been placed in it from the store over the winter, thoroughly sweep it, put up all the shelving to hold the annuals, and then get a large portion of the parking lot ready for the tree and shrub (perennials) area. It was backbreaking work, but I felt good at the end of each day, even though I was only making $15 an hour. Come May 1, and our stock started rolling off big semi-trailer trucks. M0ther's Day was one of the busiest days of the year for the centre. I toiled away, selling trees and shrubs to customers, looking after the young students who also worked with us, and constantly moving perennial plants in containers as big as 25 gallons. The work was so physical I could eat as much food as I wanted every day and never gained a pound.

I worked for the following two summers in the Garden Centre. It was fun selling plants to people who were rarely in a bad mood when they were buying new trees or shrubs to put in the garden. I had a great group of people to work with, including Tammy, Terry, who grumped a lot but had been there for years, Murray, a retired minister who would pontificate ad infinitum about the most trivial matters, including his family tree, and Larry, an old cowboy who looked after the hose fittings. Besides them, we were saddled with a group of high school and university students every summer, most of whom had no clue what hard work was. In my first summer, the centre did over $1.8 million in gross sales, the second-highest volume for any Canadian Tire store in Canada. My work at Canadian Tire was radically different from anything I had done previously in my life. It was challenging from a physical perspective. I enjoyed the interaction with the customers and helping them to understand the different plants they could buy and the growing requirements for each. I helped teach people how to plant shrubs and trees. I explained to them how the root ball had to be broken up for the plant to receive the necessary water and oxygen to establish successfully. I offered welcome advice on the sites that should be selected for certain plants. My experience as a forester, ecologist, and biologist proved invaluable in my role as garden centre worker.

I stayed at the garden centre until late August when I finally got a call from DHD to do a replacement gig for a drilling rig up in Dawson Creek, BC. I showed up at the marshalling yard in Acheson, just outside Edmonton, where I picked up my eight-person Ford passenger van and the crew. The rig workers, who were all from Newfoundland, loaded their gear and off we went. Many of the men came from other places, including BC, Manitoba, Ontario, and the Maritimes. It took me a while to get the hang of driving the oversized vehicle, which had mud tires on it, which did nothing to improve the handling. We finally made it to Dawson Creek, where I left the crew at their hotel, as I was staying at a different place. My normal routine was to get up at 5 am, pick up the crew at 6:10, then head right to the rig, which was about a 30-minute drive out of town. Once at the rig, the crew would get out, and I would wait for the night crew to load up for the drive back to town. Once they were all aboard, usually filthy dirty, I would run them back into town. The van would typically be caked with mud and strewn with garbage, which I was required to clean up. The drives were usually long and lacking in conversation, with few smiles from the crew. I would repeat this process in the evening starting about 6 pm. Besides being the crew driver, I was also expected to be the rig manager's gopher (go fer this, go fer that). Sometimes I would be asked to pick up material or equipment that was being brought to Dawson Creek by truck or travel to another town for some specific rig part, or simply obtain food or supplies for the rig manager. As often as not, I would have no errands to run, and the rest of the day between crew transfers was my own time. Essentially, I was on call for a 12 -hour period and for that, I was paid around $300 a day. There was no doubt it was good money, but on many days, I had nothing to do. I would watch TV, read a book, or just stare wistfully out the window at the snow and the trees, thinking of golf season, which was months away. Sometimes I would make pathetic attempts to burn the huge volume of calories I was consuming. Some camps had well-equipped fitness rooms, while in others the fitness room was one or two mediocre treadmill machines in a tiny, cramped area. Most camps had 'Jack and Jill' bathrooms. Two rooms shared the same toilet and shower, which at times would be quite challenging and sometimes disgusting. The routine became quite monotonous, and being away for two or three weeks at a time seemed to wear me down as the days melted into one another. The most excitement usually involved guessing what dinner would bring. The food was usually very good with a multitude of main courses to choose from, lots of different salad material, and scrumptious desserts. As time wore on, the nastiest weather of the winter set in, and the driving conditions became tenuous and extremely demanding, which didn't help matters. I had a good session in Chetwynd, BC in December. I stayed at a pipeline camp for that one, and the food was excellent as the cook had owned a restaurant in England for years. The camps had a no drinking or drugs policy.

One night a new worker for DHD showed up. Apparently, he was discovered to be drinking, and the next morning he was unceremoniously escorted out of camp. When I was in Chetwynd, the temperature dropped to -45° C. The van I was driving, a new Mercedes model, froze solid. Amazingly, there was no block heater in the van. It took two hours of charging the battery, hooked up to two trucks before we could get it started. When I called the office with my dilemma, they told me to leave it running all night. I only had one key, so that potential solution seemed ridiculous to me. It also seemed unbelievable that the rig crew could still work effectively in those frigid conditions.

I did three more jobs for DHD in January and February of 2018, then sat idle for months. Every month I was required to submit a timesheet showing that I had done nothing.

In September, I went into the office at the marshalling yard where the receptionist staffed the phone. She was not very pleasant, quite irritable most of the time, and was always unnecessarily terse with me. I handed her my timesheet, and she brusquely said, almost shouting at me, "Oh, you don't work here anymore!" I was stunned by her assertion and asked, "What do you mean?" She spat at me again, "You'll have to talk to HR."

So, I left the office and got on my cell phone right away to talk to Shorty, the head of a pathetically weak and poor HR Department. I got a phone message, so I texted him with a query as to what was going on. His cryptic answer was laughable – I'm paraphrasing. "Oh, I thought we told you. We had some complaints from operations about your performance, and there were issues with your driver reports." I wrote back, "Well, that's the first I've heard about it." I had never been advised of any 'problems' with my performance or driver reports so I was nonplussed. He said it was a done deal, and there was nothing he could do about it. I am pretty sure the operations manager didn't like me either; I didn't really fit the mold of your typical drilling company employee. All the other crew drivers came from labour backgrounds while I was an educated scientist.

As well as that, I had been sent on one job where he assigned me to a two-wheel-drive van with poor tires when all the other ones were four-wheel drive. We had got about halfway to our destination, which was a camp somewhere in the middle of nowhere west of Fox Creek, Alberta, and the weather turned really nasty. The snow was heavy, piling up rapidly, and I was driving on an industrial gravel road. On the way, we got stuck in the ditch and had to be pulled out. We finally arrived at the camp after a highly stressful trip. In the parking lot, when I was turning around, I had no control on the icy surface, and the van slid into another vehicle. I reported the incident to the camp manager and had to file an Incident Report with DHD. I never was reprimanded for it, but I was certain that the operations manager had his hand slapped. In my mind, he held a grudge against me despite it being his flawed decisions that led to me driving the van with poor tires in such bad weather, creating a potentially lethal situation.

I first thought of filing a grievance with the government for being treated so poorly but ultimately decided that there was really no point. The DHD experience had been another different and challenging work assignment. It opened my eyes up to the lengthy processes and myriad types of equipment and workers that are necessary to install underground pipelines. I got to observe first-hand the dangers and conditions that the pipeline workers were exposed to constantly, the challenging living conditions, the demanding weather environment, and the abuse that the weakest members of the crew were often subjected to.

When I got home, I decided to give Tammy at Canadian Tire

a call and see if they had any openings, and they did. I was hired the next week, working as part of the Fill Crew, which I did for the winters of 2018 and 2019. That was another different work perspective. Even though my role was that of glorified stocker, I was still privy to the machinations of a huge retailer like Canadian Tire. I sometimes had to deal with customers who needed help finding items or just simply had stupid questions. Occasionally, customers were abusive or exceptionally rude, and the standard rule was to turn the other cheek. I got to know the store layout like the back of my hand and could take a customer to just about anything that they were searching for. I was amazed at the vast array of 'stuff' we sold. I remember one old customer who said to me, "I just come in every week or two, wander around aimlessly and end up buying more junk I don't really need." That kind of summed it up.

In August of that year, Stantec hired me to complete wetland monitoring in an area north of Cold Lake, Alberta. We needed to use Argos (see below) to travel about in the saturated landscape. The work was part of a 10-year wetland monitoring program as part of CNRL's (Canadian Natural Resources Ltd.) compliance requirement related to oil and gas leases within the CLAWR (Cold Lake Air Weapons Range). The CLAWR is located on both sides of the Saskatchewan-Alberta border along the 55th latitude, covering an area of 11,753 square kilometres; 54% of the range is situated in Saskatchewan, and 46% is in Alberta[27]. With its unrestricted airspace, and state-of-the-art targets, the CLAWR is reputed to be one of the finest air training ranges in the world. It contains an instrumented aerospace testing and evaluation range, a manned air-to-ground range including a high explosive range, and an air-to-air gunnery range. DND (Department of National Defence) controls land-based access to ensure that safety, operational, and security requirements are met. Security at the site is very tight, including a two-hour orientation requirement, security screening, and formal communication protocols. It seemed weird working on a weapons range. I always wondered if we would stumble upon an unexploded ordinance and be blown to bits or mowed down by a jet with its air cannons strafing us to smithereens. On a daily basis, there was a notice issued to us, advising of areas that were being actively used and to stay away from them. As I found out, the risks weren't physical but from the strictly enforced operational protocols that DND had.

I had participated in the monitoring program in two previous years, and in 2015 I found out the hard way how strict the communication protocol was. When we were working on the range, we were always required to call the 'Sentry' (a DND communication centre) and advise them of our location whenever we left the main road to travel on a cut line. For this purpose, we were directed to carry a sophisticated radio with a portable antenna that could be placed on top of the Argo. Invariably we would be in a landscape depression, and we couldn't reach the Sentry.

On one occasion when we were unable to make radio contact, my assistant, a young archaeologist with no experience on the CLAWR, had called the Sentry on her cell phone, which I wasn't aware of. Using one's cell phone to contact the Sentry is absolutely forbidden. Within 10 minutes of her cell phone call, I heard a truck roar to a screeching halt on the road nearby. A security guard jumped out of the truck, and he started yelling at us as if he thought we were terrorists. "You have to come out to the road immediately!" We were not sure why, but we complied. Upon arriving at the road, he informed us that due to our communication indiscretion, we had to be immediately escorted from the CLAWR. On top of that, we were not allowed to return. So, we dutifully left the range and returned to our motel in Cold Lake. Ultimately, we were forced to return to Edmonton, and the remaining two crew members had to pick up the rest of our work. Within two weeks, I received a very stern, serious letter from the CLAWR Base Commander informing me of my transgression and stating that I was banned from the range for a year. In 2017 I was able to go back to the range, although I always suspected that my name was on a list somewhere marked as a former felon.

I was part of a crew of four that used Argos to drive down cut lines to the wetland sample sites. In the boreal environment, the wetlands extend for miles and miles. There is a constant array of different kinds of wetlands. Some are pure, dense, stunted black spruce stands. Others are wide open, saturated sedge fens, and there are also pockets of dry, jack pine knolls interspersed. Peat bogs are commonplace, and the depth of the peat, which has accumulated over thousands of years, is amazing. I worked on one project that involved obtaining soil samples, and we determined that the peat that we were measuring was more than three metres deep. Black flies and mosquitoes descended upon us in droves and they were merciless in their thirst for blood. Some of the wetlands had standing water up to one metre deep, and Argos were the only safe means of transportation. An Argo is a semi-amphibious, all-terrain vehicle that can travel safely across muskeg, fens, and peat bogs. The first Argos were produced in 1962, and since that time, the relatively primitive design has changed little. The machine, which has six or eight wheels with tracks, like the ones used on tanks, is driven by the rotating wheels and a one-piece handlebar steering control with a mounted brake lever. They are the most uncomfortable machines to ride in with rigid seats that lack any back support. After a day of bouncing in the Argo, I felt like my internal organs had been rearranged. They were always suffering mechanical breakdowns or getting flat tires causing innumerable work stoppages.

40 CROWSNEST PASS, ARGOS, AND ATVS

Early in the summer of 2018, I received a phone call from Clint Gellrich, a wildlife biologist who had worked at EBA when I was there. He had left them in 2015, and with a friend, he had formed a company called Beartracks Environmental, specializing in wildlife habitat work and dabbling in other environmental services. They needed someone to help with a rare plant survey in the Crowsnest Pass area of southern Alberta. I jumped at the opportunity, and Tammy at Canadian Tire was happy to let me have five days off to earn some real money. My chargeout rate as a consultant was $110 per hour, which seems like a lot, but it wasn't when compared to the rates that the big companies charged.

The Crowsnest Pass area of Alberta (and BC) has a long history, including more than 80 years of coal mining[28] (see Footnotes). CPR built the rail line through the pass between 1897 and 1898. Shortly after that, the first mine went into operation at Frank in 1900. Frank was the site of one of the worst natural disasters in Canadian history. At 4:10 am on April 23, 1903, the cataclysmic collapse of the north slope of Turtle Mountain occurred. More than 110 million tonnes of limestone obliterated a portion of the town, killing 90 people. You can drive by the site, where house-size blocks still lie in a jumbled mess. The worst mining accident in Canada's history occurred in 1914 at the Hillcrest Mine, killing 189 men. In 1979 the communities of Coleman, Blairmore, Bellevue, and Frank amalgamated to form the Municipality of Crowsnest Pass. The last coal mine closed in 1983. Interestingly that mine was at Tent Mountain, which was where Beartracks was doing the environmental work. Today, there is renewed interest in coal mining in the region, although residents are divided on the pros and cons of such resource extraction ventures.

I drove south to Calgary then turned onto Highway 22, which took me to Highway 3, where I went west to the Crowsnest Pass. The first morning of work, I drove to the starting point of our work below Tent Mountain and met Clint and the rest of the field crew. We were required to run rare plant transects (on foot) through the area being permitted by Montem Resources for a coal mine approval. My assistant was a young lady who had trouble keeping up with me as we struggled up and down the steep hills. We were in grizzly and black bear habitat, so I was making my bear warning calls on a regular basis, especially since she was a little paranoid about bears. We were both carrying bear spray, which helped to alleviate some of her anxiety. We did five days of bashing through the bush, compiling lists of all the plant species we encountered and found quite a few listed species (ones that are ranked between S1 and S3).

Plant species are ranked by Alberta Parks (BC and Saskatchewan use similar systems) according to known frequency of occurrence and rareness. The rankings are as follows: S1 - known from five or fewer occurrences or especially vulnerable to extirpation because of other factor(s), and considered the most rare or endangered; S2 - known from twenty or fewer occurrences or vulnerable to extirpation because of other factors; S3 - known from 100 or fewer occurrences, or somewhat vulnerable due to other factors, such as restricted range, relatively small population sizes, or other factors; S4 - apparently secure, the taxon is uncommon but not rare, potentially some cause for long term concern due to decline or other factors; and S5 - secure, taxon is common, widespread, and abundant. We did document quite a few rare plants, including western redcedar, western hemlock, linear-leaved scorpion weed, and male fern.

Later that summer I was asked by Stantec to participate in a rare plant survey for a proposed ATCO electric transmission line expansion. ATCO was founded in 1947 by Don Southern under the name Alberta Trailer Hire, renting fifteen utility trailers in the Calgary area. As the company's operations grew, they also began to sell trailers, first becoming the Alberta Trailer Company, then ATCO. By the early 1960s, the company had operations across North America.

The work was past Fort McKay, about 58 kilometres north of Fort McMurray. I was partnered with a regular Stantec environmental technician, and we stayed at a Syncrude camp. The camp was gargantuan, with three levels of dorm trailers stacked on top of each other, able to house more than 1,000 workers. It was like a frightening, dystopian nightmare as if the faceless oil company had turned hundreds of people into sullen-faced robots inside an immense industrial machine that was half-work camp, half-prison. Security was in full force at the camp, with guards posted at the doors to the dining room and the main entrance. The guards, who must have been brain-washed, took their job very seriously. They had batons on the hip, wore big badges and jack boots. We were not even allowed to take food items from the dining room to the lunch preparation area.

Staying there was like being in a maximum-security institution. The living routine consisted of three phases: working, eating, and sleeping, repeated on a daily basis. The workers trudged from the stacked trailers to the transport buses every day, eyes glazed over, carrying lunch bags, and wearing the requisite hardhats and full personal protective equipment. Every evening, or in the morning if they were on the graveyard shift, they poured out of the buses and shambled back to their tiny impersonal rooms. I was lucky to only have to spend a few days there, but most the of the poor sods that I saw went through the mindless routine for months, even years.

Much of work area was composed of muskeg, so we used an Argo to access the sites. As I said, I hated the Argo with a passion. I have never had an enjoyable experience while operating one. To get to the wetlands, we first had to drive the Argo on dirt roads. The routes wound their way through active logging roads, and we had to dodge logging trucks and rubber-tired skidders. As we rolled along, the Argo tracks would throw up clouds of fine dust which coated us, the machine, and all our field gear with a layer of sediment. We almost got bogged down in some of the wettest sites, but amazingly the Argo did the job without us breaking any components. Our work passed uneventfully, and I was ecstatic to say goodbye to the work camp and the Argo after a week.

In September, I went back to the Table Mountain mine site to collect ecosystem data for biophysical mapping. Our crew consisted of Clint, Mike, Bob Faye, and me. Bob was an independent soil scientist who I knew from my EBA days. The weather was crisp, with frosty mornings and sunny days. To access our sample sites, which extended right up into the alpine, we used ATVs. Unlike other jobs where the client demanded proper certification by way of completing a training course, Clint had no such requirements. He just threw me a helmet and the keys, told me to hop on, and follow him up the winding, rough track leading to the study area since we were staying in cozy cabins at the bottom of the mountain. Bob and I had absolutely no experience with the machines, which take a bit of practice to get used to. The predecessor to the ATVs (or Quads) was the three-wheeled cycle (also called ATC or tryke) which I had driven in 1981. The ATVs of 2018 were powerful, heavy machines that could be used equally well for work or recreation. However, they were dangerous if not treated with respect when used on rough terrain.

An ATV can weigh as much as 270 kilograms and go as fast as 80 kilometres per hour. They were big and fast, and it helped to be strong to control them. If you were commandeering them on sloping, uneven ground, which could be booby-trapped with logs, boulders, gullies, and erosion channels, the risk for having an accident increased. If I had known the casualty numbers before starting up Tent Mountain on the quad, I might have had second thoughts[29] (see Footnotes and Appendix).

When driving the quad, I sat astride the machine, much like riding a horse. My feet were stationary in the footrests, and I had to hang on to the handlebars, like with a motorcycle. The accelerator, which was a trigger, was controlled by the thumb on my right hand. There were handlebar brakes, also like a motorbike, and a foot brake beneath my right foot.

Bob and I were tentative the first day, but within a short time, we got the hang of the motorized beasts. By the end of the day, I was relatively proficient. For exceptionally rough terrain, like creek washouts, I found the best approach was to stand on the quad and let my legs work as shock absorbers. My field partner was Mike, who dug all the soil pits for me, and he was a speed demobn. When we were driving on smooth roads, I couldn't keep up to him. It was a constant challenge to control the mechanical beast. I had to stay focused on the task at hand at all times, trying to pick the best and safest route between the array of large rocks and rough ground. It was what I imagined riding a bucking bronco would be like. I had to make my body move with the constant up and down motion of the ATV. It was exhausting but at the same time, exhilarating. The next morning my shoulders and arms were sore from all the reefing on the handlebars, but within a few days, as I became more adept, the pain subsided.

One day we were busy working in a dense lodgepole pine forest, and I heard a branch snap. Mike was busy with the soil pit, toiling away. I said, "Did you hear that?" We were once again in grizzly country, and we had seen scat on the roads. I saw something large and brown about 20 metres away, but I didn't have a clear view of it, and I was worried that it might be a bear. With one hand on my bear spray near my hip, I slowly moved upslope to get a better look.

My heart rate had gone up, but I was relieved to see that it was a large bull moose. As soon as he saw me, he spooked and took off.

We did ten days of fieldwork, and the time went by fast. There were no bugs, the weather was pleasant, and we were in the Rocky Mountains. The views were magnificent, with verdant hills among the towering mountains, and we felt like we were at the top of the world. On the second to last day, we awoke to a cold rain driven by a strong wind out of the north. It was a full raingear day, and we knew work would be challenging and probably uncomfortable. By lunchtime, as we travelled higher on the mountain, we were in sleet, and within the next hour, it was a full-blown snow event. In my imagination, this must have been what it felt like for climbers to be scaling Everest, although I knew we were at much less risk. We did what we could to stay out of the snow but with our hands getting cold and our clothing wet, the day was curtailed early. It was a sign of fall and impending winter, and we knew it was time to go home.

41 WRETCHED CANOLA FIELDS

In 2019 I worked on a solar power project located just south of Calgary for Silverberry (Nanette Richards). The entire area of the proposed solar farm was within fields of canola. Canola oil is a vegetable oil derived from a variety of rapeseed that is low in erucic acid and high in oleic acid[30] (see Footnotes and Appendix for more information on canola). As a monounsaturated fat, it is healthier for humans than animal fats which are saturated fats. There are both edible and industrial forms produced from the seed of any of several cultivars of the Mustard family. Canola oil is also used as a source of biodiesel.

In late summer, the canola fields are in full bloom and exhibit a stunning array of brilliant, yellow fabric across much of Alberta's landscape. The lovely colour belies the true nature of the evil plant. In 2015, while I was still at Stantec, I was sent to an area south of Viking, Alberta, to conduct weed surveys on natural gas pump station sites. It was a brutally hot day with a temperature in the vicinity of 35°C. After ten hours of slogging through the endless canola, I was beaten to a pulp. My clothing was smeared with a mustard-yellow residue from the canola blossoms, my hands and arms were limp from fighting with the endless, thick stems, my mouth was parched, and my head was pounding from the heat. It was all I could do to make it to my motel and crawl into bed at the end of the day. You would think that walking through a farmer's field would be easy, but fields of mature canola were anything but that. Of all my times of thrashing through thick brush in the mountains of BC, I had never taken a beating like the canola fields dished out. The constant movement through it, which required the lifting of my feet and knees as high as I could, was debilitating and resulted in excruciating pain in the hips. By the end of the day, I could barely lift my feet.

Our job in Calgary was to check every wetland that had been previously mapped in the Alberta Merged Wetland Inventory to see if it was, in fact, a wetland. At each site, we had to confirm the wetland classification, document the plant species present, check the soils, and look for any rare species. Canola, when it is mature, can be up to one and a half metres tall, and I was not looking forward to making its acquaintance again. The plants have extremely durable stems, and invariably while traversing the fields, the stems would wrap around my feet and legs, causing me to fall forward, often flat on my face. After more than a week of struggling thru the canola and examining more than 150 wetlands, I was relieved to be heading for home.

I did a second job for Silverberry in September that involved doing TEM plots outside of Sparwood, BC. The job itself wasn't challenging. It involved three days of ecosystem sampling near one of the Teck Coal's mine sites outside of Sparwood. My assistant was Martin Robinson, a strapping, young lad who was a soil scientist with Golder in Calgary. Silverberry was working as a subcontractor for Golder. The interesting part of the project was that I was back in the southern Rocky Mountains, an area of BC that I had worked in 1975 when I was with the BC Ministry of Environment. I remember Sparwood being a small, scruffy town full of transient workers, and interestingly, it hadn't changed much in 44 years. It was always stunning and humbling to be at the foot of the immense mountains, towering over us like hundred-million-year-old geological giants crowned with distant snow and ice caps.

SECTION 8

Black Fly Environmental

The opportunity to work for Black Fly was one that I didn't expect. Jennifer Gosse, one of the owners, couldn't believe that I wasn't working for another company. I explained to her that no companies want to hire anyone over the age of 65. I believe it constituted age discrimination, but it was virtually impossible to prove. Black Fly appreciated my experience and values my contribution as a field scientist and mentor to junior staff.

42

THE TRANS MOUNTAIN PIPELINE, ROCKY MOUNTAIN GRASSLANDS, AND ARBOREAL LICHENS

In late December, I received an unexpected text from Jennifer Gosse. Jenn had worked for Stantec during my time there and had left shortly after I arrived. In 2017, she and Nadine Clifton, who had also worked for Stantec, had started a small environmental consulting company called Black Fly Environmental (BFE). The epithet was ironic for an environmental company to be named after a nasty, biting insect, but it was definitely an attention-getter. In her text, Jenn asked me if I would be interested in doing some work for them on the Trans Mountain Pipeline Expansion project. I am not sure how she tracked me down, but it was fortuitous. The work was filling the role of a Timber Clearing and Salvage Coordinator (TCSC) for Midwest Pipelines. Midwest was one of the main pipeline construction companies that had been engaged by Trans Mountain (TM). After many years of controversy and endless delays, the Edmonton to Burnaby pipeline, which was being twinned with an already existing line, was being built. A chronology of events leading up to the initiation of construction in 2019 is provided in the Appendix[31].

The expansion project has faced significant criticism, particularly from environmentalists and First Nations groups. The existing pipeline ships diluted bitumen through the Strait of Juan de Fuca, an extremely sensitive marine environment. To reach the terminus, tankers have to pass through a very narrow channel of shallow water from the open sea, with an inherently high risk of an accident and an oil spill.

Those who support the pipeline say that it will create jobs and that it has a much lower risk of having an oil spill than transporting oil by rail, which pipeline proponents say would otherwise have to be used. My own opinion is similar. As much as society continues to decrease the demand for oil products, the reality is that there will still be a need for oil and gas to be transported over long distances. In my mind, pipelines which go through intensive, regular integrity checks are much safer than rail transport of oil. I also believe that the TM Expansion Pipeline will be the last major pipeline built in Canada. The regulatory hurdles, requirements for intensive negotiations with First Nations, and immense environmental issues are forcing expenditures to be borne by corporations, higher and higher.

After the text from Jenn, my immediate reaction was negative. But the more I thought about it, the more attractive the job became. I would have to be away from home as the job involved working a six-day workweek, centred about 200 kilometres west of Edmonton in Edson, with 12-hour days. It would be extremely lucrative though, so after two days of considering the pros and cons, I got on the phone and told Jenn that I was interested. She described the job requirements, and I told her that I would be able to handle them. Once I signed a work agreement with BFE, I was required to obtain the necessary safety tickets, and then I waited until I got the call to go to work. On January 21, 2020, I drove to Edson with Annissa Robertson, a BFE environmental scientist who knew the role of the TCSC well. The next morning, I had to go through an intense four-hour orientation session with Midwest and then prep myself for the work.

On January 23, Annissa and I drove 84 kilometres from Edson to Hinton and then a few kilometres north of Hinton to where my section of the pipeline ROW clearing was being started. My job was to coordinate and monitor the clearing work on a 30 kilometres section of the Midwest pipeline construction (about 150 kilometres in total). There were four other TCSCs already working on ROW sections to the east. Annissa and I first introduced our selves to Charles, the clearing crew's supervisor, and then she went about explaining and showing me in detail what specific tasks I had to complete on a regular basis. The clearing crew consisted of machine operators who cut and stacked the trees on the ROW, processed the cut trees (delimbing them and cutting them to the lengths required by the sawmills), piled and burned non-merchantable logging debris, and mulched the cleared ROW. All the salvaged merchantable timber would eventually be transferred to the forest companies who had signed agreements with TM to accept all wood for processing. After spending the day with Annissa, I was on my own, and except for a few minor hiccups, I had no problems handling the workload.

I soon got used to the constant sound of the equipment working and learned to recognize the noise that each piece of machinery made. The feller-buckers made a high whining sound as they cut each tree stem, then a huge whump as they stacked the trees in big piles. The processors, which stripped each tree of branches, made a loud grating noise as each stem was cleaned. If the excavators were in motion, I listened for the sound of the tracks coursing over the ground and watched to see where they were moving. I always had to get the attention of the operators for a thumbs-up signal before passing them. Sometimes the operators would ask me exactly what I was doing, for as far as they were concerned, I was just some guy, with a hardhat and a high visibility vest, in steel-toed boots riding around on an ATV. I succinctly explained to them that my role was to ensure that the clearing was meeting the TM prescribed standards.

My normal work routine started off by getting up around 5:30 am, having breakfast, attending a daily safety and work issues meeting at the Midwest marshalling yard in Edson, and then driving to the ROW just north of Hinton. Once I got to the worksite, I met with Charles to see what the clearing plans and progress would be for the day. Usually, I had to unload the ATV from my pickup truck and take it down the section of the ROW being worked on to check on the status of clearing work. Unloading and loading the ATV from the truck was one of the most dangerous things I had to do, so I was always very careful with that task. There was no margin for error, but I soon became proficient at it. I monitored the logs being cut to make sure they met the required specifications and standards for the mill. This meant inspecting them for correct length, removal of defects, and being decked (stacking of logs in designated areas) properly. If I saw deficiencies, I discussed those with Charles so that he could direct the machine operators to correct them. I took numerous pictures throughout the day to document the work being done, and at the end of the day, when I returned to Edson, I selected a few pictures to be included in the daily field report which was sent to my boss. I did this work, save for one week off in February, six days a week up until the middle of March. At that time, our section was complete except for the removal of the logs, which was to be done by a logging contract company. The work schedule was demanding but I enjoyed the challenge of learning a new job and being outside, even when the weather was cold.

After the work was all done, BFE kept me on to help with proposal writing, training junior staff, and eventually to do more fieldwork in the summer and fall. That spring represented my anniversary of 50 years of fieldwork. Here I was 70 years old and still able to do fieldwork and really enjoy it. It was a prime requirement for the type of physical work we did to keep myself in good shape over the years, and I still exercise daily, without fail.

The next big field job I did was in the Rocky Mountains

in August, again near Sparwood. For that work, BFE was a subcontractor to Golder, out of Calgary. The work involved doing TEM sampling in the high elevation grasslands (HEG) of the Rockies. The grasslands were prime California bighorn sheep range, and the client (Teck Resources Limited who own and operate four massive coal mines in the southern Rockies) was in the process of reconciling the sheep ranges with the BC government, in relation to plans for mine expansions. I was part of an eight-person field crew responsible for sampling selected locations in Teck's area of interest. Every morning we loaded into a Bell 212 helicopter for the one-hour flight to our day's transect locations. The Bell 212 helicopter, which is a huge, powerful machine with a passenger capacity of 12, was used in the Viet Nam war going by the nickname of 'Hueys.' Because of COVID protocols, we were all wearing masks and gloves when in the machine, and as we were flying into the mountains every morning, I felt like I was going to war.

Our crew consisted of senior vegetation ecologist with Golder Chris Sears, vegetation ecologist Derek Rennie, contract vegetation ecologist Iraleigh Anderson, soil scientists with Golder Marcin Stanislawski, Christiane Brouwer, and Martin Robinson, contract soil scientist Dave Brown, and me. Christiane, who was always happy and cheerful, had been with Stantec when I was there, and we got along well in the field. Our helicopter drop-off spots were always on sloping ground, and the pilot would have to hold full power while we disembarked and grabbed our gear from the cargo hold. The exit only took about 30 seconds, but with the screaming turbine of the engine, the machine shaking mercilessly, the rotor wash pushing down on us, and uneven, sloping terrain, it was a stressful time. The key to safety around the helicopter was to always stay calm and to focus on the task at hand.

The steep slopes afforded ample opportunities to partake in 'trundling.' Trundling involved finding a large boulder and send it hurtling down a steep slope to see how far it would travel. It seemed to be an activity that men appreciated and participated in more than women. Perhaps it harkens back to primitive times when prehistoric men might have used large stone as implements of destruction or to kill prey. The stones on slopes in the Rockies tend to be flat and thin and not very well-suited as good trundling rocks. At one site, where Christiane and I were doing a plot, I managed to find a good trundling boulder and sent it downhill. It soon was bouncing rapidly into the air as it flew downslope, and it ultimately travelled more than 500 metres to crash into trees and brush.

For 14 days, we repeated the same routine with the helicopter drop off and pick up. The workday consisted of following a pre-selected transect route to capture TEM data at mapped polygons. The terrain was extremely steep, often on slopes of up to 80%, with scree slopes, rock outcrops, alpine ridges, and long, expansive areas of robust alpine grasses. I would invariably collect a few plants that I was not sure of, and at the end of the day, I would discuss the potential identification of them with Chris and Ira. The constant stress of walking on the steep sidehills was tough on the feet and ankles. Christiane and I were always very careful on the steep slopes. We knew that one slip could result in a nasty, potentially injurious fall. Although we knew they were there, as evidenced by the presence of scat, we never saw any sheep, and we were in prime grizzly habitat, so our bear spray was always close at hand. We saw plenty of grizzly signs, indicated by massive digging on *Hedysarum* tubers, but never had any face-to-face encounters. The scenery was spectacular, and every day we were treated to an array of stunning mountain vistas. The hiking was gruelling, but I felt privileged to be able to work in such a beautiful setting. It was stimulating and good for the soul to be out in the open, in the awesome mountains. I always felt energized and effused with a strong sense of well-being when I was in the company of them.

When I got back home from the work in the Rockies, I had a week off then headed back out into the field for another interesting project, for which BFE had been the successful proponent. I had written the proposal for this work, which consisted of collecting arboreal lichen (the ones that hang from trees) data in the Hart Mountain Ranges east of Prince George BC to estimate habitat quality for woodland caribou. Our contract was with the BC Ministry of Forests, Lands, Natural Resource Operations and Rural Development, which was quite a mouthful.

The caribou herds of the west slope of the Rocky Mountains have been significantly influenced over the last few decades primarily due to increases in industrial forestry (logging) over a widespread area east and north of Prince George. Other factors have included fires (natural), liberal hunting regulations, widespread access to snowmobile use, and predator control. Some work on arboreal lichen availability and abundance had been previously done in other areas of the province. The government wanted to try and quantify the availability and quality of arboreal lichens relative to potential snowpack depths in high-elevation, subalpine forests through the collection of accurate field data.

When I wrote the proposal, I needed an experienced wildlife biologist with a good understanding of woodland caribou and their habitat needs. As it happened, just before I had read the proposal request, I had connected on LinkedIn with Damian Power. Damian had worked with me on the Itcha and Ilgachuz, and Charlotte Alplands mapping projects back in 1997 and 1998. I hadn't had any contact with him for 22 years, although I knew he was still in the business. When I contacted him about going in on our proposal, he jumped at the chance. I thought we had submitted a good proposal, but you never knew what your chances of success would be with government jobs. The proposal request had specified a maximum budget amount of $80,000, so we knew that the successful proponent would not be chosen on price. We were pleasantly surprised to receive an email from the client saying we had been selected.

In the middle of August, I drove to Prince George where I

met Damian. He hadn't changed much in 22 years. He was still a wiry, small man with a strong Irish accent (I thought perhaps his ancestors might have been leprechauns). Damian was 75, and I was on the verge of turning 71, and between us, we had more than 100 years of field experience which was remarkable. We laughed at the thought of two senior citizens bushwhacking up slopes in the mountains, but we were tough old buzzards.

Our plan was to complete 12 transects in the mountains and gather relevant arboreal lichen along with each one. We had a limited budget, so we only had enough money for 12 hours of helicopter time. That meant we could use the chopper for doing drop-offs and pickups for four transects. The rest would be accessed by long drives of up to two hours one-way, on active logging roads to reach the mountains. We had a two-way truck radio that allowed us to set the radio frequency for different industrial roads; that way we always knew when a loaded logging truck was coming towards us. Our transects were at elevations of between 900 and 1,700 metres, with the sites with best lichen production at the higher elevations. On our first day out, we got lost and ended up doing the first transect in a site relatively close to Prince George, which gave us a chance to get our sampling routine figured out. I was responsible for completing an ecosystem plot with a list of plant species and associated percent cover estimates. While I was doing that, Damian rated the arboreal lichens that were hanging from the tree branches on site. He had to give a productivity rating from 1 to 5 (with 1 being the highest and 5 having no lichen cover). The arboreal lichens consisted of two kinds, *Bryoria* and *Alectoria*. By the end of the first day, we had a cohesive, efficient routine down pat. Our sample sites on each transect were spaced approximately 50 metres apart. We were in a mixed stand of subalpine fir and Engelmann spruce, as were most forests on our transects.

The second day presented us with a different ecosystem, one dominated by my old friend and nasty foe, devil's-club. Over the rest of the field trip, we encountered various conditions, including subalpine meadow and open forest complexes, dry ridges, and subdued, forested slopes. The weather stayed fine until the last three days when we encountered cool, rainy, fall conditions. On one of those days, we had a long drive, deep into the Hominka River valley. It was cool, and the clouds, heavy with rain, were hanging just above the treetops. Our access road to our selected transect location was blocked, and we ended up walking the last two kilometres to get to the starting point. The road was sloppy, and in the mud, we saw grizzly, black bear, moose, and deer tracks. The entire study area was prime bear habitat, and we were on full alert that day.

We had seen two small black bears and a beautiful, young bull moose on the roads before the last day of our trip. On that morning Damian and I had walked up through a cut block into our forest to be sampled. At our first stop, I started making my list of plants when Damian whispered, "There's a bear." I thought he was joking and said, "Sure." He laughed quietly and said, "No, look over there." I looked to where he was pointing, and sure enough, a large black bear was watching us from about 30 metres away. He had probably heard us and had come over to investigate. My heart started pounding. He sat stationary when we yelled at him. Thankfully, he was a well-behaved bear, and he turned and ambled away. Within a few minutes of seeing the bear, we heard a bull moose grunting. It was rutting season, and he was trying to attract the interest of any females in the area. We never saw him, but we heard him continuing his overtures for a while.

The fieldwork had been fun, and I enjoyed Damian's company. The work had been physically challenging, but it was a challenge that we both willingly took on. It was gratifying to know that we could still participate in the fieldwork exertions that we both enjoyed so much. It seemed just a few years ago that I had been a young buck embarking on my wilderness career, but here I was, 50 years later, still able to do what I loved. We parted ways in Prince George, and I headed back to recover from 30 days of fieldwork over a 37-day period.

At the beginning of October, I was sent with Curtis Vieville, another professional biologist with BFE, to carry out a wildlife sweep and ecosystem sampling for a proposed gravel pit near High Level, AB. High Level is about 740 kilometres north of Edmonton. Our study area was in a decadent (full of decay) forest of trembling aspen. The wildlife sweep, which was done by Curtis, was a regulatory requirement to ensure that no animals, including birds, were living or nesting within the proposed work area. I did the ecosystem sampling. The forest was quite open and filled with thickets of prickly rose. They were the biggest and nastiest rose bushes I had ever seen, with some specimens being over two metres tall. Luckily, our work only took three days, but after I got home, I counted more than 40 rose thorns in one hand, and I was wearing gloves while working. I just never knew what fieldwork would bring.

43 BACK IN THE ROCKIES

In the spring of 2021, when the leaves were barely emerging, I received an email from one of my bosses asking if I would be willing to participate in the continuation of the HEG program. Golder was planning to launch a massive field program running from mid-June into September. They were looking for vegetation ecologists and soil scientists to complement their own staff and form a field crew of 24 people to carry on with the TEM work that had been initiated the previous year. I jumped at the chance and told my boss to sign me up for the entire summer. I knew from last year that the work would be arduous and challenging, but the opportunity to go back to the southern Rockies was too good to pass up.

Work was planned to begin on June 15, with the schedule broken into ten-day shifts and four days off between shifts. Because most people could only work for a few shifts (or even only one or two), a total of 51 scientists participated in the sampling program with only myself, Nick Brown, Martin Robinson, Danielle Mai, and Mathew Turnbull working on all six shifts.

For the entire work program, we were stationed at the Elkford Lodge work camp just outside of Elkford, BC. Due to COVID restrictions, the camp was in lockdown mode all summer. That meant we had to be masked at all times in camp, were not allowed to socialize and had to eat our meals alone in our tiny rooms. On top of that, we were required to be rapid-tested (for COVID) three times during our eight days during each shift in camp. We nicknamed the camp – Elkatraz.

On June 14, 24 scientists and the Golder Project Manager assembled at the camp where we all had to participate in the project orientation. The next morning, 12 crews of two people departed from camp to learn how to complete the TEM plots on iPads and to become familiar with the requirements of the work program. The first two days started out slowly with a bit of a gong show, which was not surprising with organization and logistics to ensure that all people had the required equipment and necessary understanding of the comprehensive data collection standards.

By the third day (June 17), we had been organized into three cohorts of eight people. Each cohort would fly out together in the Bell 212 at the beginning of the day (spaced one hour apart) and be picked up later in the afternoon. Dave Hawrys, of Ascent Helicopters, was our pilot for the summer, and he turned out to be fearless and very good. He had to be, given that most of our landing and pickup spots would be on steep slopes, which required toe-in landings. Our field days were approximately ten hours long, leaving two hours for the evening work of identifying and pressing collected plant specimens and reviewing and editing the field data. Christiane was my partner for the first shift, and we renewed our acquaintance as we hiked through subalpine meadows with carpets of emerging spring flowers, including yellow glacier lily, white western pasque flowers and dainty spring beauty. Deep areas of snow were still present in landscape depressions and under dense forests.

On June 20, Christiane and I, as well as Hearthy Majordo and Sandra Marken, all got out of the chopper on a steep snowbank at the top of an alpine ridge. Dave was holding full power as we climbed out of the machine on what turned out to be slippery, melting snow. Hearthy and Sandra, both with limited helicopter experience, looked terrified, but I had a firm grip on both of them. We huddled near the front of the helicopter as we waited for Dave to lift off. Once he had departed, I looked at where we had gotten out and realized that if one of us had slipped, we would have jettisoned down a steep snow slope to certain injury. In hindsight, we should have looked for another landing site.

Because of the time of year and the fact that a significant amount of snow was present at high elevations, the rest of our transects for the first shift were relatively easy. This gave me a chance to get used to the thin air and for my legs to become stronger with each day.

For my second shift, which began on June 28, I was paired with Justin Dyck, a geologist from Golder. At the age of 36, he could walk my ass off, but we developed a good work routine and got along well together. Justin carried trekking poles (he didn't need them because he carried the shovel, which could be used for support), and he kindly lent one to me, which was of huge benefit when traversing down steep slopes. Going downslope was much more difficult than going uphill. Gravity was always my enemy going downhill. Uphill, although harder on the lungs and heart, was so much easier because you could kind of lean into the steep slopes. When I returned home to St. Albert, I bought my own set of trekking poles.

Justin Dyck, my field partner for three shifts

The weather had started to heat up, and daytime highs, even in the mountains, exceeded 30 degrees C. It meant we had to increase our intake of fluids during our strenuous days and add electrolytes to the water we were drinking. By July 1, southern British Columbia was officially experiencing a heatwave with temperatures in some parts of the interior exceeding 40 degrees C. On that day, Lytton recorded a new all-time high temperature for Canada of 49.6. The temperature in the Elk Valley was over 37, and with us working on steep, hot slopes, we were also exposed to extreme temperatures. We had three successive days of excessive heat, and the entire crew was suffering from the effects. I was consuming about four litres of water a day in order to stay adequately hydrated.

On July 5, we awoke to light rain, which gave us a brief respite from the heat. Because the valley was socked in with low cloud, we were forced to work from our trucks which meant collecting data from roads at relatively low elevations. I was driving the truck up a very narrow, overgrown track to the starting point of our transect when we had a mishap. Due to the thick shrubbery along the edge of the road, I had no way of seeing the right-hand road base. Suddenly we heard a thump and scraping sound in the right wheel well. We got out to look and saw that two large branches protruding from the shrubs about 30 centimetres above the ground had jammed into the crease between the wheel well and door. The door was somewhat crumpled and didn't open properly, but we were able to continue down the road and complete a meagre number of plots. I was upset at causing the 'accident,' but nothing significant came of the incident, although the damage cost about $5,000 to have repaired.

On July 6, the weather cleared, and the oppressive heat began to build again. Thankfully our shift ended that day, and on July 7, I drove back to St. Albert to have four days of rest before beginning shift three.

For that shift, my partner was Hearthy Majordo, a young Golder reclamation scientist from Vancouver. She had never worked in the mountains or done TEM work before, so her first shift at the beginning of the program was a steep learning curve for her.

On July 13, her first day in the field with me, we were working on a steep rubble slope. Around ten o'clock, I noticed two people approaching us from downslope. My initial reaction was that they were from a different crew in our group, working in the same general area. Then I suddenly realized that the woman was wearing shorts and the man was clad in a singlet. They were not part of our group but were two young people working at an Elk Valley bible camp who were just out for a leisurely jaunt up a death-defying mountain. They were interested in what we were doing, so I spent about 20 minutes giving them a brief synopsis of our work program and how we collected data. We said goodbye to them and watched as they quickly scaled the slope above us and disappeared over an alpine ridge.

On July 17, the heat was still overbearing, with temperatures daily exceeding 32 degrees. We started our transect on some classic high-elevation grasslands, then travelled to a steep forested slope that we had to traverse down to our pickup spot in the valley bottom. The stand was composed of continuous, dense, young lodgepole pine. The going was tough, and after about 30 minutes, I heard Hearthy yell behind me. She had fallen and twisted her ankle – not severely, but bad enough that she was slowed down even more. So, we gradually picked our way down the steep slope for a distance of three kilometres through the endless pine trees. Eventually, we arrived at our pickup spot exhausted and soaked in sweat.

Very fresh grizzly bear digging

That evening Hearthy iced her ankle, and the next morning before work, she wrapped it with a tensor bandage. She was able to walk on it, but she was definitely hobbling. Martin, who selected transects for each crew the night before, assured us that the transect would be 'easy' so that Hearthy would not be overly challenged. We were dropped off high on an alpine ridge, and our transect appeared to be relatively benign. For most of the day, we meandered our way down towards our pickup spot, a talus slope below us. I travelled slowly and never got too far ahead of Hearthy, ensuring that she was able to cover the rough terrain without too much trouble. Ultimately, we arrived at a steep rock-strewn gully that was a formidable barrier. We took our time and picked our way down the slope without any falls. The 'easy' transect turned out to be quite a bit harder than we thought.

Our last day of the shift offered more demanding terrain and heat, but our pickup spot was near a creek with a bouldery floodplain. We arrived at the site about an hour ahead of pickup time, so we doffed most of our clothing and took advantage of the cold water to soak our weary bones. The rushing, crystal-clear water was stimulating and soothing and provided a welcome respite from the sun.

For the fourth shift, Justin was returning to the program, and I asked the project manager if we could be paired together again. We had jelled as a team, and it just made life simple for both of us to work together without having to get used to a new partner. By this time in the summer, mid-July, there were numerous forest fires burning to the west in southern BC. Smoke was being blown into our study area, which made conditions difficult and challenging.

Indian hellebore with steep cliffs and talus in the background

On our first day of fieldwork, July 27, we were all issued respirator half-masks to help with breathing if the smoke became too dense. Justin wore his for the morning and had to discard it at lunchtime. He said that when going upslope, the mask made getting a sufficient supply of air difficult. I never did use my mask either on the fourth shift or the fifth one, as I never encountered any symptoms from the smoke. Some field crew members suffered from sore throats and burning eyes.

Whenever the opportunity arose, Justin and I would trundle rocks down steep slopes. It was one of the few times in the mountains while we worked that we could relax and enjoy short breaks. Our best trundling site would be encountered during shift six, the last one of the field program.

On many of our days, we were treated to an array of spectacular mountain vistas, and only occasionally were our transects in relatively boring terrain. On August 1, we did a transect that consisted primarily of herbaceous avalanche tracks, which were favourite sites of the resident grizzlies. One of our sample sites was dominated by Indian hellebore, purple-hued aster and lush, vigorous cow-parsnip. It appeared that a backhoe had dug up much of the cow-parsnip, but it was only the extensive digging of a grizzly that had caused large areas of soil to be exposed. The bears focus on in the roots and typically will revisit the sites with ample food supply on an annual basis. The digging of the roots stimulates new growth to ensure an endless source of vegetation, almost as if they were 'gardening' them. The extensive digging made the hairs on the back of my neck stand up, although the evidence appeared to be at least a week old.

The rest of the shift passed uneventfully, although the presence of smoke continued to present challenges. On August 2, fieldwork was cancelled for the day because the valley was so full of dense smoke that Dave was unable to fly the helicopter from Fernie to the Sparwood airport to pick us up. August 3 was our last day of fieldwork, and on the fourth, I headed back to St. Albert, driving through the dense layer of smoke. It seemed as though the coming of the apocalypse was upon us.

For the fifth shift, my field partner was Polly Uunila (nee Stokes), a terrain expert from Squamish, BC. I hadn't seen Polly since 1994 when she worked with me on two contracts for the BC government. At that time, she was a young, impressionable university grad; now, she is a seasoned terrain mapper working independently. We got along well and spent time catching up on our lives as we worked. The smoke from the fires was still causing problems but was not bad enough for us to use masks. On August 15, the smoke had become so dense that a number of workers refused to work, although Polly and I stuck with it. We were one of only two crews who completed the entire day of work. The project manager decided that due to the smoke conditions, the shift would be shut down a day early, and we de-mobilized on August 16.

When I returned to St. Albert, I visited the dentist on August 19 to have all of my remaining top teeth extricated and replaced with dentures. I had a full top plate and partial lower plate. Needless to say, my gums were extremely sore and tender, and my diet for the immediate future would be primarily a liquid one. The last shift at camp would be a challenge finding enough soft foods that I could comfortably eat.

Author doing steep slope grassland plot in smoky conditions

The first field day of the last shift was August 24, and the valley was finally free of the never-ending smoke. On top of that, the temperature gradient had shifted to a cooler regime with overnight lows down to 4 or 5 degrees and daytime highs reaching only 27. It was beginning to feel and look like fall in the mountains. I had started work in the middle of June, and now, just over eight weeks later, the vegetation was showing signs of the end of the growing season. Many plants were turning yellow and brown, and typically flowering was finished, with only dried seed heads remaining. We called the act of identifying dried and shrivelled plants 'forensic' botany.

On August 26, Justin and I had a chilling near-encounter with a grizzly. We had been working our way down a series of avalanche chutes and were doing a plot when we heard a blood-curdling sound. It was a powerful roar combined with a loud growl. I have no doubt it was a grizzly, and we estimated it to be about 200 metres away in another avalanche chute, one that we had planned to travel down. Given that we didn't know exactly where the bear was or which way it would be travelling, we changed our sampling route to avoid it.

Over the course of the summer, there had been five or six close calls with black and grizzly bears, but no one had ever been in danger or had to use their bear spray. Martin and his partner had experienced the closest event, hearing a large grizzly rummaging in the forest only about 75 metres from where they were working. They quickly and quietly moved away from the potential threat without any face-to-face incident.

On August 27, we found ourselves on a 'perfect' trundling slope. We were hiking down a 75 percent slope which was consistent for about 700 metres downhill. The boulders perched on the incline presented ample opportunities for prime trundling. The ultimate trundle was a large boulder, about 60 centimetres across, that Justin was able to initiate its downward race by using his feet to release it from the hold of gravity. Within a few seconds, that boulder was bouncing into the air and ultimately travelled about 500 metres downslope, crashing into other rocks and coming to rest in an extensive shrub patch far below.

A late-summer, early morning in the Rockies

The rest of the shift passed quickly, and finally, on August 31, the field sampling program came to an end. I was sad to say goodbye to the incredible and stunning Rocky Mountains, but it had been an amazing summer working in the most spectacular environments. It was a privilege and a pleasure to be constantly challenged both mentally and physically. I had survived the entire summer without any injuries, save for a few scrapes and bruises. It was amazing that with a total of approximately 1,000 person-field days that not one person had been seriously hurt while travelling over the endless treacherous terrain.

Epilogue

So, there you have it, a rambling account of my 51 years of fieldwork in western Canada. I am now 72 and have no intention of stopping.

I have always wondered where my deep-seated passion for nature came from. When I was a wee lad in England of only two or three, my parents lived in a small caravan in the forest. I was allowed to roam freely through the surrounding woods, and I am sure that must have made deep, long-lasting impressions on my psyche. Later, in Canada as a child, and up until I was an adolescent, we used to ramble over hill and dale. TV was limited to two black and white channels, and there were no electronic devices available to distract us. We would think nothing of hiking up Hollyburn Mountain or walking across the Lions Gate Bridge to explore Stanley Park. During those formative years, I was a cub and a boy scout.

In Boy Scouts, we often had wonderful camping trips into the wilderness where we were free to act like savages and 'rough' it. It was stimulating and fun to build a campsite, make a functional fire pit, chop down trees, and listen to the sounds of the forest and rushing rivers. So, I have no doubt that all those adventures taken together sowed the seeds for my future career development and employment.

As Damian said, "As long as I can put one foot in front of the other, I will keep on doing fieldwork." Fieldwork never gets dull. I am always in a different place, with different ecosystems and new scientists. I am more than happy to share my knowledge with budding young resource technicians and professionals. There is always something to see and learn that I have never encountered before and that stimulation helps keep my brain and body young. I have had the privilege of working in some amazing environments with some incredible scientists. I will continue to hike through the forests and into the mountains to document our wilderness areas.

I figure that one day they will have to pry the shovel from my cold, dead hands.

Now it is autumn. All the leaves have fallen from the trees, and the blanket of white has descended. I can hardly wait until next field season to see where I go and what wonders await me.

Helicopter Stories

Over my career, I have probably spent more than 1,000 hours flying in a variety of different helicopters. Without these amazing whirlybirds, we would not have been able to access the wilderness areas where we often worked. The pilots who flew the choppers were an amazing breed, able to land in the tightest spots, negotiate swirling winds, and navigate mountainous terrain. Some of them were eccentric; some were talkative, others quiet and thoughtful. It is expensive to earn a helicopter pilot's license, with the cost today varying from $50,000 to $86,000 depending on the model. I have included some specific incidents involving helicopters and pilots in the chapters of this book, but this section is an expansion of that, with some additional anecdotes.

Up until the 1960s, soil surveys in BC were done on horseback. Following that, pickup trucks were used to access the backcountry. In the 1970s, helicopters made an appearance in fieldwork projects.

The first time I ever flew in a chopper was in 1975 in the East Kootenays. The machine was a Bell 47, which was first produced in 1946. Early models varied in appearance, with open cockpits or sheet metal cabins, fabric-covered or open structures, some with four-wheel landing gear. Later model D and Korean War H-13D and E types settled on a more utilitarian style. The most common model, the 47G introduced in 1953, could be recognized by the full 'soap bubble' canopy, exposed welded-tube tail boom, saddle fuel tanks, and skid landing gear. The later three-seat 47H had an enclosed cabin with full cowling and monocoque tail boom. This was the model seen in the TV series 'MASH.'

Our pilot, Neil, was the eccentric type. He sat in the middle of the tiny cockpit while my partner, Mike Fenger, and I sat on either side of him. I remember us coming up on a grizzly running through a young, open forest and was amazed at how fast he could run. Sometimes when Mike and I got out at a stop to do a plot, Neil, in shorts and T-shirt, would go for a quick run up a steep mountain slope. Other times he would work on the sweater he was knitting.

My next helicopter experience was during the Northeast Coal project in 1976. Our pilot was Bill Carr-Hilton, and the machine was a Bell JetRanger 206. The 206 had a top speed of about 220 km per hour; today a new one sells for $1.2 million. The overall design has changed little since 1966 when they were first produced commercially. The first machines we flew in had flat windows, and it was hard to see anything, such as the forest below us. Later, probably in the early 1980s, Bell replaced the flat window with a bubble version which allowed us to look straight down into the forest, wetlands, or the mountains as we were flying. Bill, a happy man with years of experience, was a very good pilot. He could always judge where the wind was coming from and land in the smallest, little holes. Pilots were not supposed to drink at night before flying the next day, but Bill would often join us playing cards while he sipped on a gin and tonic. It never seemed to affect his flying abilities. One day when we were on the way home, I was sitting in the back with Wayne Blashill and Maureen Ketcheson. Wayne, who was in the middle, had fallen asleep. I leaned forward and tapped Bill on the shoulder, at the same time motioning towards Wayne. Bill knew exactly what to do. He took the machine up in elevation, then suddenly dropped it out of the sky. Poor Wayne woke with a start, his eyes bugging out of his head. I am sure he thought we were going to crash.

In 1977 I was doing a reconnaissance flight in the Kimsquit River valley with one of the soil scientists from our office. We had landed in a small wetland that appeared to be dominated by peat moss. I was in the back, on the same side as the pilot. Without thinking, I jumped out of the chopper and promptly went straight down, up to my chest in cold water. I never made that mistake again.

We used a helicopter occasionally on the Canstar project, north of Fort McMurray, in 1981. That was where the camp was a gong show, and we were all behaving like clowns. The pilot's name was Steve. He wore a cowboy hat when he wasn't flying and used to wear a 44 Magnum pistol on his hip when we were doing bush work. His machine was a Hughes 500, which looked like a flying egg. It could fit three passengers in the back, but it was very tight and uncomfortable. I remember him demonstrating a hammerhead turn. He would only do it when there was no one else in the machine with him. The Hughes 500 was light, with a top speed of close to 240 kilometres per hour. He would take the chopper up high, then come screaming down towards the ground and make a tight turn, going back up then rolling it upside down in a steep arc until it was the right way up again. It was a very impressive stunt but only strengthened our opinion that he was a little crazy.

In northwestern BC in 1982, again with a JetRanger 206, I had the thrill ride with Paul the pilot through the Grand Canyon of the Stikine. I am sure there were not too many pilots who would have repeated his act of screaming through its length at top speed. To this day, it still has been the most exciting heli ride I have ever been on. Also, in 1982 Denny and I had our trip to the Tashsish-Kois drainage on the west coast of Vancouver Island. Our pilot Pete, also in a JetRanger, picked us up one day from our camp at the mouth of the Tahsish River. It was an overcast day with the clouds low in the valley. We didn't think that Pete would be able to take us halfway up the mountain slope above us to drop us off in a small wetland, but he did. We couldn't see a thing, and neither could he, but we followed the tree line up the slope until we reached our drop off site. We got out, and he soared off into the fog. Pilots are not supposed to fly in conditions like that.

In 1985, when we were flying from Meziadin Junction in northwestern BC, our pilot was Gary, who liked chasing animals. One day when we had flown a few kilometres from camp, I heard him suddenly yell, "F***!" Whenever I heard a pilot swear, I was immediately concerned. As he swung the chopper around in a tight turn, he said, "We have to go back, right now." It turned out that the battery, which was housed in the nose of the machine, had malfunctioned. The battery had melted down and could have blown the front right off the machine.

When Dave Yole and I were doing the Observatory Inlet work, we were flying the coastline one day when the pilot noticed a helicopter standing in tidewater near a beach. He was curious about why the chopper was in the water, so we circled around and landed nearby. We walked to where the pilot was standing on the beach. His story was this: he had landed when the tide was out, waiting for his crew to arrive for a pickup. He didn't realize it, but he had left his lights on, and when he returned to the machine to start it up, the battery was dead. Unfortunately, the tide came in and flooded the machine. I heard that he was fired due to his error in judgement.

I knew a biologist by the name of Brian Fuhr who had a terrifying helicopter experience. They were flying somewhere in northern BC and were going down to land on a small tabletop mountain. The pilot descended much too fast and hit the rocky promontory hard. The helicopter bounced up into the air and started to autorotate. For a helicopter, 'autorotation' refers to the descending maneuver in which the engine is disengaged from the main rotor system, and the rotor blades are driven solely by the upward flow of air through the rotor. The freewheeling unit has a special clutch mechanism that disengages any time the engine rotational speed is less than the rotor rotational speed. If the engine fails, the freewheeling unit automatically disengages the engine from the main rotor allowing the main rotor to rotate freely. When the machine bounced up, the body of the chopper started to spin faster than the rotors. As the machine smashed down onto the mountain again, it started to come apart. Brian said the rotors flew off, and at one point, the seat (with the passenger in it) next to the pilot was facing him. The helicopter finally came to rest in a shattered state, hanging on the side of the mountain. They were lucky that no one was killed or seriously injured. He said that the pilot walked away from the crash and never flew again. Brian was shaken up, but later in life went on to become the Regional Wildlife Biologist in Smithers, BC.

I never came close to crashing, although I do remember one time in the Itcha and Ilgachuz Ranges in 1997, our pilot had a warning light come on showing an engine malfunction, and he landed immediately. He tried to figure out what was wrong with the machine with no luck. Eventually, he took off again, with us aboard, saying it would be a rough ride. I recall the chopper wobbling into the air and barely clearing the trees at the edge of our landing spot. He ended up dropping us off at camp and going back to his base to fetch a replacement machine.

In 1991, when Brent and I stayed in the Snip Gold Mine camp in the Iskut River drainage, we had Tom as our pilot. He was a giant of a man. Brent remembers him as being fastidious about having immaculate windows, cleaning them at almost every stop, and doing clean and jerk exercises with large logs while he was waiting outside the helicopter. Brent also remembers, on another project, having an ex-Viet Nam pilot who he thought was going to be a good chopper operator. Turned out he was wrong. The guy needed huge areas to land in, and the crew ended up nicknaming him 'Tarmac John'.

And then there was Chuck, who was the scariest pilot I ever flew with. The last straw came when we were coming in for a tricky landing on an alpine ridge, and he turned to Larry Lacelle and said in a panic, as his eyes bugged out, "Which way is the wind blowing?" We had him replaced in short order.

When Brent, Denny, and I worked out of the guide outfitters camps in the Kechika River and Muskwa River valleys in northeastern BC (1994), our pilot's name was Rob. There was nothing wrong with his flying, but he used to spend nearly an hour in the bathroom every morning. When he emerged, every hair on his head was neatly in place, and his clothing was pressed and immaculate.

When we started doing the MacBlo projects on Vancouver

Island, we were lucky to be able to use the services of the Forest Industry Flying Tankers. There were four excellent pilots who flew the LongRanger (206L) choppers. The machines seated six passengers and had more power than the normal 206 versions. I remember one pilot whose name was Bill Hill, nicknamed Hillbilly Bill. He was an older guy with a shock of grey hair and massive beard. He had a great sense of humor, was fearless, and could land the machine just about anywhere.

Later when we did projects in Haida Gwaii, our regular pilot was Rob. Rob was a tad overweight and barely fit into the pilot's seat, but he knew the islands like the back of his hand. He was obsessed with finding glass balls on the outer coast. Glass floats, glass fishing floats, or Japanese glass fishing floats are popular collectors' items. They were once used by fishermen in many parts of the world to keep their fishing nets, as well as longlines or droplines, afloat. Large groups of fishnets strung together, sometimes 80 kilometres long, were set adrift in the ocean and supported near the surface by hollow glass balls or cylinders containing air to give them buoyancy. These glass floats are no longer used by fishermen, but many of them are still afloat in the world's oceans, primarily the Pacific. Once a float landed on a beach, it might have rolled in the surf and become etched by sand. Many glass floats showed distinctive wear patterns from the corrosive forces of sand, sun, and saltwater. When old netting broke away from a float, its pattern often remained on the surface of the glass where the glass was protected under the netting. Other floats had small amounts of water trapped inside of them. This water apparently entered the floats through microscopic imperfections in the glass while the floats are held underwater by netting. To accommodate different fishing styles and nets, the Japanese experimented with many different sizes and shapes of floats, ranging from five to 50 centimetres in diameter. Most were rough spheres, but some were cylindrical or 'rolling pin' shaped. One day, when we were flying on the outer coast, Rob spied a large glass ball trapped in driftwood near a rocky beach. He promptly found the nearest landing spot, jumped from the helicopter, and took off like a man possessed. We waited, and in about 20 minutes, he returned proudly holding the giant glass float, sweating profusely with a huge smile on his face.

Last year I had the new experience of flying in a Bell 212 helicopter in the Rockies. The pilot, Craig, was very good. He could put us down in the toughest spots even when it was windy. The machine is very powerful and has a top speed of 223 kilometres per hour. For all of the HEG work this year we again used the Bell 212. Our pilot Dave, was amazing and could hold power as we disembarked or climbed aboard at the roughest of locations.

I never get tired of flying in helicopters. They are the fastest and most efficient way to access sample sites, see the landscape, and cover a huge amount of territory in a short time. It is always cool to look down as we cruise over the infinite array of vegetation patterns, rugged mountain vistas, and endless water features.

Preparation for Fieldwork

Getting ready for fieldwork is challenging and a very important part of project planning. Ideally, when you arrive at your base for fieldwork (whether it is a hotel or camp), you should have everything you might need as well as backup supplies. If you are staying in a camp, you may be a long way from any stores where you can buy a needed item. Following is a description of what one should normally take with them and consider for different kinds of work, based on what I have learned over 50 years.

Backpack or Cruiser's Vest

When you are crashing through the bush, there are two options on how to carry all your field gear: a backpack or a cruiser's vest. I have never worn a cruiser's vest, preferring to have my implements, field items, and safety gear in a backpack. The cruiser's vest has lots of pockets for books and tools, plus space for other things, such as a compass, GPS unit, or clinometer (for measuring slopes), which can be attached to it for easy access. It is entirely a matter of personal comfort and how much gear you have to carry. If your pack or vest is overloaded, making it excessively heavy, the day can be long, tiring one.

What to Carry

Everything you need for the day's work has to be with you. This includes a notebook, field guides, bug repellant, sunscreen, measuring devices, tools, safety items (including a small personal first aid kit), water and food. The pack or cruiser vest will quickly get heavy, so one has to be prudent about what to take.

Clothing and Footwear

What you wear is an important consideration, depending on the type of work being done, the kind of environment you are working in, and what the weather will be like. You need to be comfortable and also protected from the elements. Modern-day, light pants and shirts are great if you are not doing real bushwhacking. If you are fighting through thickets of obnoxious plants, like devil's-club or roses, you will need clothing that will provide more protection. I recommend thick, cotton (or cotton blend) work pants. I have used them for years. They are relatively light, dry fast when wet, and protect your legs. NEVER, ever wear jeans. They are the worst kind of pants for real bushwork. When they get wet, they bind on your knees and take forever to dry. If you are working in hot, dry environments (such as grasslands on the prairies), light cotton clothing is the best. For working in cold weather, layering is the best approach. Consider wearing a light cotton undershirt (usually with long sleeves), a cotton work shirt, and a wool top. Stanfield makes the best version of this. Remember, if you get too warm, you can discard layers; if you are cold and don't have enough clothing, it can not only be uncomfortable but dangerous. In the winter, a pair of lined work pants are great; some people prefer wearing long johns under work pants.

It is not worth compromising on footwear. Cheap boots will not stand up to the daily wear and tear, and you cannot risk getting blisters if you are walking substantial distances or up and down steep slopes. There are many different types that are designed for specific types of work. For summer fieldwork, a high-ankle hiking boot is suitable, although for many types of work, it is mandatory to have boots with a steel toe that are CSA (Canadian Safety Association) approved. For working around wetlands, rubber boots do the job, but they are not so good for walking long distances because they have no ankle support. For winter work, an insulated boot is absolutely necessary. Dunlops are great.

Four types of field boots:

- Rubber boots with steel toes
- Knee-high winter boots with steel toes and removable

insulating liners
- Construction boots with steel toes
- Regular hiking boots with ankle support

Jackets and Raingear

As with boots, it is not a good strategy to buy cheap outerwear. A strong, light jacket will serve you well. Goretex jackets are excellent for rainy weather, but they tend to be very expensive. When I worked on the coast, I always had my rain gear in my backpack, even on sunny days. As nice as the weather could be at the beginning of a workday, by the afternoon, you could be engulfed in rain. Better to buy good quality raingear, for it will keep you drier longer and be more durable over time. On the coast, my raingear would be worn out and replaced every two years of fieldwork.

Gloves and Head Covering

Gloves are handy, especially if the bush is thick or the bugs are bad, although it is difficult to use equipment while wearing them. In the winter, lined gloves are a nice addition. Headgear comes in many types and varieties, but they are another necessary item for fieldwork. In the summer, I have always worn a ball cap, although for work involving pipelines or construction, it is a mandatory requirement to don a hard hat. I never wore one in the bush, finding that it was always falling off. Some people like a hat with a wide brim to help protect against the sun, and for work on the prairies, I wore a wide-brimmed straw hat.

Vehicles

For most projects, you need a truck. Sometimes you can get away with a standard pickup truck if you are working in the vicinity of a town or city. For real bush work, a 4x4 truck is necessary. If you are renting a vehicle, make sure it has all-weather radial tires or even off-road tires. Check to make sure the truck has a functional jack and an inflated spare tire. You will have to add the following items to the truck: spill kit, first aid kit, flares, safety pylon, fire extinguisher, axe, mattock, and 'piss can' (a plastic backpack unit with straps and a spray nozzle that is only needed during fire season). A roll of paper towel and sanitary wipes come in handy as well. When driving trucks on the backroads, it pays to take your time and constantly stay focused on the road ahead. If the road begins to look sketchy (lots of mud with big areas of water, deep ruts, or trees across the road), it pays to stop and check out the conditions before proceeding. It's not worth it to get stuck, especially if you are a long distance from any help or if there is no cell phone coverage.

ATVs and Argos

ATVs and Argos are occasionally needed to access project areas. Both machines are designed to get you where you need to go. Make sure that you have the necessary training to drive them and, above all, respect the power and risks that both present. For an ATV, you should carry a small fire extinguisher, a spill kit, and a few bungee cords. They are not as prone to breaking down as Argos, but you still need to be careful how you operate them. Argos are renowned for having different components fail, but as with the ATV, you should carry a fire extinguisher and spill kit. It goes without saying that you need a certified safety helmet and good gloves to operate both machines. When driving the Argo in long grass, be careful and watch for logs or boulders that can be hidden. I knew a young, inexperienced driver, who when going down a steep, grass-covered slope, slammed into a rock. The impact threw her passenger into the windshield, shattering it. Luckily, he was wearing his required helmet.

Helicopters

When you arrive at the helicopter base or the chopper comes to pick you up, make sure that the pilot does a helicopter orientation with you and your partner or crew. Even if he doesn't offer one, ask him to do it. It will allow you to understand what he expects from you for entering and leaving the machine, how to stow your field gear, safety requirements, and communication protocols. If you are carrying bear spray, it will have to be stowed in a screw-top canister which must always go into the cargo hold. When you are approaching the chopper, ensure that any long items, primarily a shovel, are held horizontally at waist height. If the pilot is holding power as you exit, he will usually want you to crouch beside the machine where he can see you. Typically, you will need to give a thumbs-up sign to let him know that it is okay for him to leave. As he departs, you or one of the crew should call him on your hand-held radio to do a radio check with him. If the pilot is okay with you walking away from the machine when he lands, make sure you always walk away on his side of the machine, either perpendicular to him or slightly ahead of the machine. Usually, he will only let you walk away if the ground is level or near level. You must NEVER walk towards the back of the machine. When you are waiting for a pickup, and the machine is coming in for a landing, make sure you are hanging on to any headwear. In the process of getting into or out of the chopper, the most important thing is to stay calm and cool. The whole process takes but a few seconds, but with the turbine engine howling and wind from the rotors rushing by you, it is easy to feel stressed. So be calm and stay as relaxed as possible.

Survival Kit

Heli drops for fieldwork are when the pilot leaves you at one location, and you arrange for him to pick you up at another location. When you do these, it should be standard practice to leave a small survival kit at the pickup spot. You should not assume that the machine will always return. In all my years of helicopter work, I have never had to spend a night in the bush, although I have come close. Years ago, Dr. Bob Ogilvie and Adolf Ceska were doing botany work on a small island somewhere off the coast of BC. The pilot left them in the morning with plans to pick them up at the end of the day and flew back to his base. The only problem was that he left to go on holidays and forgot to tell anyone that they were to be picked up. Poor Bob and Adolf spent the better part of three days waiting for a chopper that never came. No survival kit, no sleeping bags, no extra food, and no hope. I am not sure if they even had a lighter or matches with them. The kit should contain a few important items: a lighter and safety matches, space blankets, granola bars, water, non-perishable food, basic first aid kit, string and rope, fishing line, flashlight, mirror, candles, survival knife, and a whistle. Oh yes, and toilet paper. I always carry some in my backpack. You never know when nature will call.

Biting Insects

Biting insects, including mosquitoes, blackflies, deer flies, horse flies, and no-see-ums often make life miserable in the bush. Sometimes they can be prolific even in places where you least expect them. I have experienced clouds of mosquitoes on the prairies and swarms of black flies on top of unvegetated mountains. Having some form of bug repellant is necessary, although applying too much of the chemical can wreak havoc with your skin. I prefer the spray cans over liquid or cream applications. A bug hat (netting that just fits over your head) is a really good idea, although it does impede one's vision to some degree. Some people like to wear a full bug jacket which includes a head covering, although I have never used one. Whatever you decide to use to combat biting insects, make sure you have something.

Wasps and Bees

Invariably, during a field season, you will have an encounter with a wasp or bee nest. Sometimes while walking, you or your partner will inadvertently kick a nest hidden in the ground or in a decaying log. As soon as you hear frantic buzzing, it is time to run away! And run away fast! If anyone on your crew has an allergy to bees or wasps, make sure that they have an epinephrine pen with them. It is also a good idea to carry a few antihistamine pills.

Bears

For anyone doing fieldwork away from urban areas, having Bear Awareness training is likely mandatory. Having said that, I can offer a few pointers that might be useful. You can count on running into black bears anywhere in Canada that has uncleared forest and brush. When you are in bear country, it pays to follow these basic rules: watch for bear signs (digs, stumps torn up, boulders overturned, tracks, mark trails, mark trees, day beds, and scat), and if you think there are bears about then make noise (you never want to surprise a bear). Keep your eyes and ears open (but don't be surprised if a bear sneaks up on you; they can be remarkably quiet) and keep your bear spray handy (if you have a belt, attach it close to your hip – on the side of your dominant hand). If you don't have a belt, then attach it to your cruiser's vest or backpack, where you can access it in an instant. If you run into a sow with cubs, try not to position yourself between the mother and cubs. If you are attacked, you must fight for your life. Never run from a bear. If the bear is threatening you, then back away slowly, trying not to make eye contact. All the training information says to play dead if you are attacked by a grizzly. I am not so sure about that. I still think I would fight for my life.

Sunscreen

In the summer and fall, sunscreen is an essential item. Pay special attention to your ears, nose, and the back of your neck. Having a sunburn is no fun and, if bad enough, can make you quite sick. In hot and sunny weather, make sure you apply it before starting fieldwork and reapply at regular intervals throughout the day.

Water

Never take having water for granted; DON'T EVER forget it. Being dehydrated is dangerous and unpleasant; the input of water throughout the day is an important but simple exercise. My policy is to always carry two litres with me. I know it is heavy, but as you drink it during the day, your pack or vest will be getting lighter. Two litres is the absolute minimum if you are working in hot weather. If temperatures exceed 30°C, you might need even more. In cool weather, it is easy to get complacent and forget to drink water, but if you are exerting yourself, it is an important thing to remember. You cannot rely on finding a fresh, reliable water supply when you are in the field. In most of Alberta and Saskatchewan, there are no good sources of fresh drinking water. If you are in the mountains, you can often find fast-moving water streams. However, you should be aware of the risk for contracting Giardia (also known as beaver fever). It is a small parasite that lives in slow-moving, back-country water. The symptoms of Giardiaisis, which may begin to appear two days after infection, include violent diarrhoea, excess gas, abdominal cramps, upset stomach, and nausea. Resulting dehydration and nutritional loss may need immediate treatment. A typical infection can be slight, resolved without treatment, and last between two to six weeks, although it can sometimes persist for longer and be more severe.

Food

You will need to have some form of food and nutrition with you. What people take to provide them with energy varies tremendously. There are endless options. Personally, I never eat much when I am working. If I am doing strenuous hiking or digging soil pits and have too much food in my tummy, I get quite uncomfortable. I like to eat a protein-rich breakfast and have a big dinner. I typically don't take much in the way of snacks. Some people simply snack all day. Trail mix, dried fruit, chocolate (not too much), power bars, and vegetables are all good choices. I usually have a fish (salmon or tuna) sandwich around lunchtime, as fish is easier to digest than meat. I haven't eaten ham, cheese, or chicken in the bush for many years. I had a bad experience with a turkey sandwich years ago. I had purchased one when we landed in the chopper near Meziadin Junction. I ate it quickly then we took off. Within a few minutes, I was overcome by nausea and had to ask the pilot to put the machine down, where I promptly vomited. Turkey spoils faster than any other kind of food. Just a thought.

Communication

Communication is critically important, especially if you are working in remote areas away from cell phone coverage. Hand-held radios are necessary to talk to the helicopter pilot and potentially other crews. If you are planning to drive on active logging roads, you shouldn't even attempt it without the benefit of a truck radio. It is folly and exceptionally dangerous to do otherwise. These days, crews can carry an InReach unit which allows message sending from remote areas to a central location. At times you may need Satellite radios to make contact with your office or the client.

Electronic Devices

Tablet computers are certainly remarkable machines. I first used one on the solar farm project in Calgary in 2019. You can load apps for GPS, mapping (Avenza), determining UTMs and elevation, and take site photos. At the end of the day, you can upload your data and photos to an off-site storage location or send them to a client. They can do a lot, but they do have their limitations. If it's raining, the touch screen isn't particularly responsive; in bright sunshine, there is significant glare from the screen that makes it hard to read information. In cold temperatures (less than -20°C), they stop working completely. If you are using a tablet, make sure it is fully charged every night, and if you are using it a lot during the day, you will need to carry an extra battery pack with you. For GPS units, they also need to be fully charged, or if you are using a unit that uses batteries, make sure you are carrying an extra supply of batteries. Up until the last few years, I always recorded my data on hard copy field sheets. Standardized field data sheets on write-in-the-rain material are available from Crown Publications (Victoria, BC). Even today, with the advent of electronic devices for data recording, I still keep a supply on hand for emergencies.

Soil Pits

You may be lucky and have an assistant or a soil scientist with you who is responsible for digging your soil pits. It is not always a requirement to do a soil description, depending on the kind of work you are doing; if you are doing an ecosystem plot, then it is required. For completing a reliable soil description and determining the type of surficial deposit you are in, the standard rule is to dig down until you reach un-weathered parent material (between 70 to 100 centimetres). It may not be possible to dig that deep. Often in soil material, you encounter limiting factors, such as compact basal till, cementation from chemical processes, or excessively stony soils. In those cases, you will have to do your best.

If you are working on roads, a road cut (in a ditch or on a roadside cut bank) can be a big help to figure out what is going on. Sometimes you will have to dig the soil pit yourself or help someone else dig it. Either way, there are a few good rules to follow. If your hands are not tough, wear gloves until you develop calluses;

otherwise, it is easy to get blisters. Make sure you have a good reliable shovel and always carry an extra one in the truck or back at camp. Never buy one with the flanges at the top of the shovel blade (they tend to get caught on roots as you pull the shovel up from the pit). It is easy to break a shovel if you or your partner is not careful. You should carry a trowel or soil knife to help you pick away at the soil profile when you are trying to decide what it is, and also for doing hand textures.

Some people use saliva to do hand texturing, but I rarely do that. I find it disgusting. I usually empty out a dish detergent bottle and fill it up with water despite adding more weight for the pack or vest. You need water not only to do hand textures but also to clean your hands when you are done.

You should also carry a small carpenter's tape for measuring the soil horizons. Soil material is excessively hard on the hands, so it is important to have hand lotion to apply to your hands every night before bed. Clay material is the worst – it literally sucks the moisture from your skin. If you don't apply lotion, your hands will end up extremely dry and even cracked.

If you are digging a soil pit on a slope, always pile the soil from the pit on the upside of the hole. It is much easier to fill in the pit when you are done. If you are working in deciduous stands, expect lots of roots at the soil surface. You can carry a hatchet, but they are very heavy. A better option is to keep the shovel blade sharp for cutting through the roots; if the roots you encounter are large, then simply move the digging location. A soil auger comes in handy for describing wetland soils and also for getting deeper in soil pits of primarily fine-textured or sandy material.

A lot of things to consider as you prepare for another field trip. As you do more work, you will likely develop your own routine for what to take and how to pack it. Whatever you do, make sure you are comfortable, protected, and safe. Oh yes, a couple of last items. If you are doing soil pits, make sure your fingernails are trimmed down and keep a nail brush back at base. If you are walking on steep slopes, make sure your toenails are cut short.

Typical items in the backpack:

- Leather winter gloves and light summer gloves
- Rain pants
- High visibility vest
- Bear spray, screw-top canister
- Bear spray
- 2-litre metal water-bottle
- Silva compass
- Carpenter's tape measure
- Personal first aid kit
- Increment borer
- Putty knife
- Toilet paper
- Empty dish detergent bottle for soil-texturing water
- Safety glasses
- Extra pens, pencils, and felt pen markers
- Toque for cold weather
- Ball cap for regular field days
- Plant identification field guidebook
- Field notebook

Note that mosquito repellant and sunscreen are not in the picture but would normally be included.

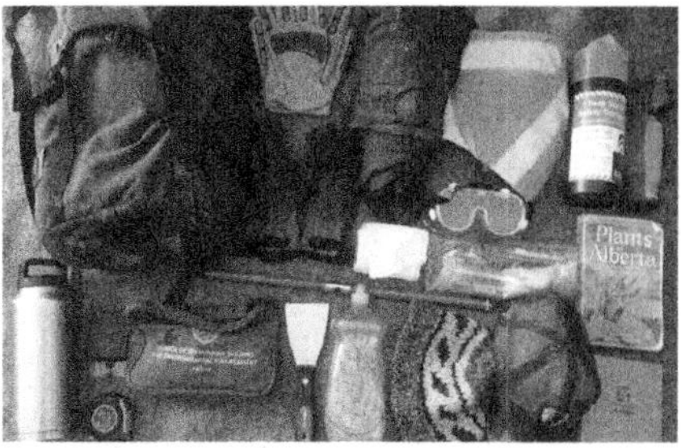

Glossary

annual allowable cut

AAC - the designated amount of timber a license holder (timber supply areas and tree farm licenses) can cut on an annual basis; established every ten years.

arboreal lichen

Lichens that grow on trees, primarily hair lichens that hang from branches.

azimuth degree

An angular measurement in a spherical coordinate system (the angle between true north and a point of interest).

biogeoclimatic classification

The classification of ecosystems based on vegetation, soil or terrain, and macro-climate; developed by Dr. Vladimir Krajina in BC in the 1950s and adopted by the MOF as the foundation for forest management.

biophysical inventory

A program of collecting data on vegetation, terrain, soil, and wildlife.

bog

A wetland typically dominated by black spruce and/or peat mosses fed by nutrient-poor surface water.

carrion

The carcasses of dead animals.

choker man

The person who puts chokers (a steel cable with a slide on the slip hook and lock) on the logs so they can be pulled out of a cut block.

climax forest

A forest dominated by trees that represent the final stage in the process of ecological succession.

cryoturbation
The mixing of soil materials due to a reoccurring cycle of freezing and thawing.

ecosystem classification
An organizing system that defines delineated portions of the landscape (in BC called site series; in Alberta and Saskatchewan called ecosites).

ecosystem plot
Data collection, including vegetation, terrrain, and soil, at a selected point in the landscape.

Environmental Impact Assessment
A formal assessment of the potential impacts to all environmental components relative to a proposed resource extraction or infrastructure project.

epiphytes
Plant species, such as ferns and mosses, that establish and live on other species (typically trees).

estuary
A partially enclosed body of coastal brackish water (mix of saline and fresh water) subject to tidal influences and constantly receiving sediment from the rivers that flow into them; they typically have marsh vegetation and are highly productive.

fen
A wetland typically dominated by sedges and shrubs fed by mineral-rich groundwater.

floodplain
A flat area of land with a stream or river running through it that extends from its channel banks to the edge of the enclosing valley (typically with sandy or silty soils).

fluvial
Moving water.

genera

The plural form of genus, a primary taxonomic category ranked between species (below) and family (above).

glaciofluvial

Waters flowing from melting glacial ice that transport gravel, sand, and silt.

heavy metal uptake

The well-documented process by which certain plant species draw heavy metals up from the soil and concentrate them in the plant tissues.

increment borer

A tool, consisting of a handle and long hollow tube with a boring end (like a screw), that is used to bore into a tree; a tray is pushed into the tube which allows a tree core to be removed. The annual growth rings of the tree can then be counted.

insectivorous plant

A member of a group of plants specializing in the trapping and digesting of insects by way of specialized plant structures.

muskeg

A broad term loosely applied collectively to the wetlands of the Boreal Region, including peat bogs, sedge fens, water-logged marshes, and shrub swamps.

old-growth

Generally considered to be forests that have never been logged (>200 years old on the coast and >180 years in the interior).

parent material

Unmodified soil or geologic material.

pedologist

A soil scientist, someone who describes soil profiles.

pioneer tree species

A species of tree that repopulates disturbed areas; they are usually light-demanding, deciduous, and with their dense canopy allow shade-tolerant evergreen trees to establish.

plant ecologist

A scientist who studies the distribution of plants and the interactions between the plants and the environment.

plant taxonomy

The identification, description, classification, and naming of plant species.

prescribed burning

The use of fire as a tool to modify existing vegetation and produce or maintain a more desirable type of cover.

second-growth

A regenerating crop of trees growing on a site after it has been logged.

seral forest

A forest that develops following natural disturbance (including fire, insects, and windthrow).

surficial material

Material that has been deposited on the earth's surface by some physical process or action (including water, gravity, movement of ice, wind, or volcanic eruption).

symbiosis

From the Greek 'living together' is any type of close and long-term biological interaction between two different organisms (includes mutualism, commensalism, and parasitism).

tafoni

Weathered sandstone bedrock, characterized by patterns resembling Swiss cheese.

talus

Piles of rocks and large boulders that accumulate at the base of a cliff or slope.

telemetry

The process of tracking radio-collared animals (usually ungulates or predators) with radio receiving equipment.

terrestrial lichen

Lichens that grow on the ground, including soil, rocks, and logs (includes crustose, foliose, and fruticose [and others]).

Trapper Nelson

A backpack with a frame invented in 1920 by Lloyd Nelson.

vascular plants

Those plants that contain conductive tissues (xylem and phloem).

Footnotes

1 – Information on the Japanese internment camps is modified from Slocan Extension Internment Camps – on the Heritage BC website (2020)
https://heritagebc.ca/japanese-canadian-location/slocan-extension-internment-camps/

2 – The species account of the wolverine is modified from the Wikipedia website (2020)
https://en.wikipedia.org/wiki/Wolverine

3 – Historical data on Anyox is modified from the Wikipedia website (2020)
https://en.wikipedia.org/wiki/Anyox

4 – Logging fatalities in 2005, from the CBC News website (2005)
https://www.cbc.ca/news/canada/b-c-logging-industry-deaths-more-than-double-in-2005-1.539894

5 – Bear bile Explained is modified, from the National Geographic website (2019)
https://www.nationalgeographic.com/animals/reference/bear-bile-explained/

6 – Kermode bear - Wikipedia
https://en.wikipedia.org/wiki/Kermode_bear

7 – BC hunting stats are from the Guide Outfitters Association of BC website (2016)
https://www.goabc.org/economic-impact/

8 – The chronology of the creation of the Gwaii Haanas National Park Reserve, National Marine Conservation Area Reserve, and Haida Heritage Site is from the Parks Canada website (2019)
https://www.pc.gc.ca/en/pn-np/bc/gwaiihaanas/info/histoire-history

9 – The species account for the common raven is modified from Wikipedia website (2020)
https://en.wikipedia.org/wiki/Common_raven

10 – The description of the raven in Haida mythology is modified from the Wikipedia website (2020)
https://en.wikipedia.org/wiki/Haida_mythology

11 – The species account for the rubber boa is modified from Reptiles of BC on the Thompson Rivers University website (2004-2020)
https://www.bcreptiles.ca/snakes/rubberboa.htm

12 – The description of Haida Watchmen is modified from Watching Over the Past: The Watchmen of Haida Gwaii on the Reader's Digest Canada website (2019)
https://www.readersdigest.ca/travel/canada/haida-gwaii-watchmen/

13 – The demise of the Golden Spruce is modified, from Kiidk'yass, the Golden Spruce on the AtlasObscura website (date unknown)
https://www.atlasobscura.com/places/kiidk-yaas-the-golden-spruce-2

14 – Details on Chris Czajkowski and her account of a bear attack are taken from Diary of a Wilderness Dweller (Czajkowski, C. 1996 [second edition published in 2005 by Harbour Publishing Co. Ltd., Madiera Park, BC]).

15 – The species account of the Clark's nutcracker is modified, from the Wikipedia website (2020)
https://en.wikipedia.org/wiki/Clark%27s_nutcracker

16 – Information on The Long, Long Trailer is modified, from the Wikipedia website (2020)
https://en.wikipedia.org/wiki/The_Long,_Long_Trailer

17 – Accounts of traditional use of native plants are modified after Plants of Southern Interior British Columbia (Parish,

R., R. Coupé and Dennis Lloyd [editors]. 1996. BC Ministry of Forests and Lone Pine Publishing, Vancouver, BC.

18 – Information on insectivorous plants is modified from Wikipedia websites for the pitcher plant, common bladderwort, and sundew (2020); information on the common butterwort is taken from the NatureGate website (no date provided)
https://en.wikipedia.org/wiki/Pitcher_plant
https://en.wikipedia.org/wiki/Utricularia
https://en.wikipedia.org/wiki/Drosera
http://www.luontoportti.com/suomi/en/kukkakasvit/common-butterwort

19 – The species account of the prairie rattlesnake is modified from the Wikipedia website (2020)
https://en.wikipedia.org/wiki/Crotalus_viridis

20 – The description of the Elbow River floods is modified from the City of Calgary: Calgary Flood Story website (2020)
https://floodstory.com/history

21 – Information on the history of Cassiar is modified from the Wikipedia website (2020)
https://en.wikipedia.org/wiki/Cassiar,_British_Columbia

22 – The description of jade mining is modified from the Wikipedia website (2020)
https://en.wikipedia.org/wiki/Jade

23 – The chronology of the Coastal Gas Link protests is taken from the Wikipedia website (2020)
https://en.wikipedia.org/wiki/2020_Canadian_pipeline_and_railway_protests

24 – The description of the Arctic research project is taken from Shrub Monitoring in Canada's Arctic Using Multi-Scale Measurements from Field Plots, Unmanned Aerial Vehicles and Satellite Remote Sensing (produced by Clement, C. and R. Fraser. 2017. For Polar Knowledge Canada, Ottawa).

25 – The account of traditional use of Inuit carrot is taken from Flora of the Canadian Arctic Archipelago (Aiken et al. 2011);
https://nature.ca/aaflora/data/www/faoxnma.htm

26 – Account of the Fort McMurray fire in 2016 is modified from the Wikipedia website (2020)
https://en.wikipedia.org/wiki/2016_Fort_McMurray_wildfire

27 – Information on the Cold Lake Air Weapons Range is modified from The Encyclopedia of Saskatchewn on the Canadian Plains Research Centre, University of Regina
https://esask.uregina.ca/entry/cold_lake_primrose_lake_air_weapons_range_4_wing.jsp

28 – Details regarding mining in the Crowsnest Pass are modified from the Wikipedia website (2020)
https://en.wikipedia.org/wiki/Crowsnest_Pass,_Alberta

29 – ATV statistics are from the MotoShark.com website (2020)
https://www.motoshark.com/atv-accidents-injuries-statistics/

30 – Information on canola and rapeseed is modified from the Wikipedia website (2020)
https://en.wikipedia.org/wiki/Canola_oil

31 – The chronology of the approval process for the Trans Mountain Expansion Pipeline is modified from the Wikipedia website (2020)
https://en.wikipedia.org/wiki/Trans_Mountain_pipeline

Appendix

Contained herein is expanded information for sections that were briefly discussed on the identified pages.

Page 15 – In Slocan City, there were 595 Japanese Canadians in the camp at the end of 1942. As with other internment camps located in small BC towns, people were housed in old hotels, decrepit buildings, and mine houses. With over 1,800 people, Lemon Creek was the largest of the nine West Kootenay internment camps constructed during World War II. Over 1,000 Japanese Canadians were interned at the government-leased farm of Emilie and Konstantine Popoff during World War II. This was the last internment camp built, completed in 1943. Nearly 100 buildings were erected, both small cabins and large dormitories. The large bunkhouses were for work crews and later adapted for the elderly, bachelors, and single people.

Bay Farm held 1,376 Japanese Canadians by the end of 1942. Temporary tents were set up, and then shacks were built in rows. Along with Lemon Creek and Popoff, this was one of the major camps in the lower Slocan Valley. Famous Japanese Canadians who were interned at the Bay Farm included environmentalist David Suzuki and writer Joy Kogawa.

Page 32 – **Wolverines** are stocky, strong animals. With short legs, a broad and rounded head, small eyes, and short compact ears, they more closely resemble a bear than other members of the weasel family (Mustelids). Although their legs are short, they have large, five-toed paws with crampon-like claws and plantigrade posture (walking on the soles of the feet, like a human or bear), which enables them to climb up and over steep cliffs and towering mountain peaks with relative ease. An adult wolverine is about the size of a medium dog, with a length usually ranging from 65–107 centimetres and a weight of 5.5–25 kilograms, though exceptionally large males can weigh up to 30 kilograms. They have thick, dark, oily fur, which is highly hydrophobic, making it resistant to frost. This has led to the popularity of their fur among hunters and trappers as a lining for jackets and parkas. Like many other mustelids, they have potent anal scent glands used for marking territory and sexual signaling. The pungent odour has given rise to the nickname 'skunk bear.' They possess a special upper molar in the back of the mouth that is rotated 90 degrees towards the inside of the mouth. This special characteristic allows them to tear meat from prey or carrion that has been frozen solid. Most of the wolverine's sustenance is derived from carrion, on which it depends almost exclusively in winter and early spring. They may find carrion themselves, feed on it after a predator (often, a pack of wolves) has finished, or simply take it from another predator. Wolverines are also known to follow wolf and lynx trails, purportedly with the intent of scavenging the remains of their kills. Prey mainly consists of small to medium-sized mammals. Still, the wolverine has been recorded killing prey such as adult deer and other ungulates, many times larger than itself.

Page 63 – Anyox. Granby Consolidated started construction of the town in 1912. Within a few years, Anyox had grown to a population of almost 3,000 residents, as the mine and smelter were put into full operation; rich lodes of copper, gold and silver were mined from the nearby mountains. Coal to fuel the smelter was shipped from the company built and owned town of Granby on Vancouver Island and Fernie in southeastern BC. Anyox had no rail nor road links to the rest of British Columbia, with access served by ocean steamers that traveled to Prince Rupert and Vancouver. The company town was a large operation with onsite railways, machine shops, curling rink, golf course, and a hospital.

In the spring of 1918, Granby Consolidated built the first wooden tennis court in Canada. That same year, incoming ships brought the Spanish flu epidemic to Anyox. Dozens of workers and residents died from that epidemic. In the early 1920s, concrete pioneer and dam engineer John Eastwood designed a hydroelectric dam which, at 48 metres high, was the tallest dam in Canada for many years. The town was almost wiped out by forest fires in 1923, but the townsite was rebuilt, and mining operations continued. Acid rain from the smelter stripped the foliage from trees on the hillsides, which soon became bare.

The Great Depression drove the demand for copper down, which spelled the beginning of the end for Anyox. Operations held on for a short time but were steadily scaled down while the company stockpiled 45,000 tonnes of copper, three years of production, that it could not sell. The mine eventually shut down in 1935, and the town was abandoned. Salvage operations in the 1940s removed most of the machinery and steel from the town. Forest fires in the 1940s burned the remaining wooden structures. During its 25 years of existence, the mines and smelters produced 4 tonnes of gold, 230 tonnes of silver and 340,000 tonnes of copper.

Page 80 – **Bear bile** has been used in traditional Chinese medicine for thousands of years, with the first reference appearing in an eighth-century medical text prescribing bear bile for illnesses like epilepsy, hemorrhoids, and heart pain. In the early 1900s, scientists discovered that bear bile, a fluid that is produced by the liver and stored in the gallbladder, contains a significant amount of ursodeoxycholic acid—more than other animals. This acid is medically proven to help dissolve gallstones and treat liver disease (information on bear bile and raising bears in captivity is taken from the National Geographic website [2020]). Bear bile, however, is also marketed as a cure for cancer, colds, hangovers, and other conditions, although there is no scientific evidence supporting such claims. It is also sometimes used an ingredient in household products like toothpaste, acne treatment, tea, and shampoo as a way to expand the market for bear bile beyond traditional use.

Nowadays, the Chinese keep bears in captivity to provide a reliable source of bile. China began farming bears in the 1980s. While farming was intended to take pressure off wild bears being poached for their gallbladders, many consumers prefer bile from wild bears;

they believe it to have more medicinal strength. Nonetheless, today thousands of bears are kept in cages for this purpose, primarily in China but also in Vietnam, Laos, and Myanmar.

Asiatic black bears, also known as moon bears, sun bears, and brown bears, are some of the most common species farmed for bile. They have a life expectancy of 20 to 30 years in the wild, but they can survive up to 35 years in captivity, meaning the poor animals can spend decades at a farm. Bear-bile farming has been widely condemned for being inhumane. The bears are often kept in cages so small they cannot turn around or stand up, and neglect and disease are commonplace.

Page 81 – Kermode bear. It is the official provincial mammal of British Columbia. While most Kermode bears are black, between 100 and 500 fully white individuals exist. The white variant (spirit bear) is most common on three islands in British Columbia (Gribbell, Princess Royal, and Roderick), where they make up 10–20% of the Kermode population. The Kermode bear was named after Frank Kermode, former director of the Royal B.C. Museum, who researched the subspecies and was a colleague of William Hornaday, the zoologist who described it. White Kermode bears are not albinos, as they still have pigmented skin and eyes (Kermode bears can also be all black).

Genetic studies found that white Kermode bears breed more with white Kermode bears, and black Kermode bears breed more with black Kermode bears, in a phenomenon known as positive assortative mating. One hypothesis is that this happens because young bears imprint on their mother's fur colour. Kermode bears are omnivorous for most of the year, surviving primarily on plants and berries except during the autumn salmon migrations when they become predators. White bears have more success than black bears in capturing salmon. Salmon evade large, black bears about twice as often as they evade white bears, giving white bears an advantage in salmon hunting. The white fur of the bear is harder to spot by fish underwater than black fur is, so the bears catch fish more readily.

Page 98 – Unsustainable logging plans initiated a public proposal to protect the South Moresby Wilderness Area. Protests, blockades, and other civil disobedience happened on Lyell Island, but logging continued in the face of legal and political controversy until July of 1987 when Canada and British Columbia signed the South Moresby Memorandum of Understanding. One year later, the South Moresby Agreement was signed, paving the way to Canada's designation of the area as a National Park Reserve. The Government of Canada and the Council of the Haida Nation signed the Gwaii Haanas Agreement in 1993. The Agreement expressed respect for Canadian and Haida interests and designations and included a mutual commitment to protect Gwaii Haanas. Under the agreement, the South Moresby archipelago officially became the Gwaii Haanas National Park Reserve. The official name change of the Queen Charlotte Islands to Haida Gwaii (Islands of the Haida people) was enshrined in 2010.

Page 102 – Ravens in Haida mythology. The raven is a trickster – a figure found throughout Indigenous cultures around the world. While tricksters are constructed very differently from culture to culture, they share similar characteristics; namely, they are most often foolish and childlike troublemakers. Some tricksters can be harmless, others can be heroes, and still others even cruel or selfish. Another common thread among tricksters is that they are wanderers, both physically and spiritually. They often travel between the spirit world and the tangible world, as well as the areas in-between. Some, like the raven, can shape-shift to become human, spirit, animal, or even inanimate objects.

In Haida culture, the raven is believed to be the most powerful of mythical creatures. His appetites include lust, curiosity, and an insatiable desire to interfere and change things and to play tricks on creatures of the world. It is important to note that while raven is an important part of how Haida see the world, he is not thought of as a god per se. He symbolizes creation, knowledge, and prestige and the complexity of nature.. He also represents the unknown and demonstrates that every person sees the world in a through different eyes. He is also a healer and keeper of secrets (description of the raven is modified from The Raven in Haida Culture [Raven Reads 2018]).

Page 106 – Species account of the rubber boa. Its loose skin and many small, smooth scales combine to give the snake its rubbery appearance. Adult boas are dark olive green to brown ish colour, and their bellies are pale yellow to orangish-yellow. As juveniles, however, their small size and pinkish colour are strikingly similar to that of an earthworm. However, the rubber boa is much more fascinating than its simple exterior suggests. Also known as the 'Two-Headed Snake', it can be difficult to tell a boa's head from its tail. Rubber Boas have little to no neck and their eyes are tiny. When being attacked, the snake contributes to the confusion by curling in a ball and hiding its head beneath its body. The tail, equipped with a small hard cap, is then presented as a decoy, and the snake often will jab its tail about as if striking. This way, the snake avoids any harm to its vital head region. If the predator remains undeterred, boas can release a smelly musk from their vent. However, that's as much as this sluggish, secretive snake can muster – rubber boas never bite and are happy staying safe beneath a damp log or rock.

A true boa constrictor, the rubber boa belongs to a famous family of snakes. This family includes, among others, reticulated pythons and anacondas. As proof of this relationship, the rubber boa has tiny vestigial (remnant) limbs on either side of its vent. These tiny 'spurs' are more evident on males than females. Boas usually are crepuscular (active at dawn and dusk) and nocturnal, and they function well at these cooler temperatures. Perhaps due to their sedentary, sluggish lifestyle, female rubber boas may reproduce once every four years. However, they can live for up to 30 years.

Rubber boas eat infrequently, even compared to other snakes. When they do catch prey, however, boas use two main tactics to obtain a typical meal of mice or shrews. When preying on adult rodents, boas constrict their prey, wrapping their body around the prey until its heart stops. When preying on baby rodents, boas use their tail as a club to keep the mother rodent out of the nest while they swallow the nestlings. In fact, many boas are heavily scarred on their tails from the bites of mother mice. Rubber boas are not limited to hunting at ground level. These snakes are able swimmers, climbers, and burrowers, enabling them to vary their diet. Other food that boas eat include bird eggs, fledglings (baby birds), nestling rabbits, lizards, other snakes, salamanders, small chipmunks, and bats (information on the rubber boa is paraphrased from Reptiles of British Columbia [Thompson Rivers University 2020]).

Page 115 – Haida Watchmen. For generations, Haida posted sentries over their villages. These watchmen would raise an alarm when danger was in the offing. Several decades ago, the Haida Nation became concerned about vandalism of sacred sites with widespread logging underway and unregulated tourism. In 1981 people volunteered to watch over several sacred sites, using their own boats to reach remote beaches. The watchmen, who are all from the Haida Nation, are trained in wilderness first aid, marine emergencies, small vessel operations, bear awareness and cultural awareness, among other things. The watchmen are a cultural contact for visitors and also assist in any emergency situations that arise. From late spring to early autumn, two or three watchmen are stationed at Windy Bay (Hlk'yah GaawGa), Hotspring Island (Gandll K'in Gwaay-aay) and Anthony Island (SGang Gwaay Llnagaay) as well as at Tanu (T'aanuu Llnagaay) and Skedans (K'uuna Llnagaay) (Tanu and Skedans lie outside the reserve but within the Haida Heritage Site). The prime responsibility of the watchmen is to guard and protect the natural and cultural heritage of these sites. While the totem and mortuary poles of the villages gradually decay and the house beams are left to fall back to the earth, these modern-day Haida watchmen maintain a link to a bygone era.

Page 129 – Clark's nutcracker. All Clark's nutcrackers have a sublingual pouch (in the floor of its mouth and in front of the tongue) capable of holding around 50–150 seeds, depending on the size of the seeds. The nutcrackers store the seeds, usually in the ground for later consumption, in caches of one to 15 seeds. Depending on the cone crop, a single nutcracker can cache as many as 98,000 seeds per season. The birds usually store more than they actually need as insurance against seed theft by other animals (squirrels, chipmunks, etc.), as well as low availability of alternative foods; this surplus seed is left in the cache, and if not utilized, may be able to germinate and grow into new trees, if the conditions are right. Through this activity of caching and over-storing, the bird is perpetuating its own habitat. Closely tied in with this storage behaviour is the bird's remarkable long-term spatial memory; they are able to relocate caches of seeds with great accuracy, as much as nine months later, and even when the cache sites are buried under deep snow. These memory powers are another example of the sophisticated intelligence of Corvids. Seeds are taken both from the ground and from trees, where the nutcrackers are very agile among the branches. The birds can extract food by clasping cones in such a way that the cones are held between one or both feet. The birds then pry the cones open with their strong bills.

Page 166 – Praire rattlesnake. The prairie rattlesnake commonly grows more than one metre in length. In Montana, specimens occasionally exceed 120 centimetres. They carry the distinctive triangle-shaped head and pit sensory organs on either side of the head. They are found in North America over much of the Great Plains, the eastern foothills, and some valleys of the Rocky Mountains, from southwestern Canada through the western United States to northern Mexico. In Canada, they occur in BC, Alberta, and Saskatchewan. Habitat characteristics can vary depending on subspecies and specific locale. Generally, prairie rattlesnakes occupy areas where an ample supply of food species live.

They have also been known to occupy burrows of other animals, and they are typically found in dry, rocky sites that are warm and sunny. They sometimes climb into trees or bushes and will even rest in crevices and caves. They are typically active diurnally (active during the day) when the weather is cool and nocturnally during hot weather.

These snakes, armed with powerful venom, use about 20-55 percent of their poison in one bite, and will defend themselves if threatened or injured. As with other rattlesnake species, prairie rattlesnakes will rapidly vibrate their tails, which produces a unique rasping sound to warn intruders. Because of their wide geographic range, the snakes have a wide choice of menu options. Generally, this species prefers small mammals, such as ground squirrels, mice, rats, small rabbits, prairie dogs, or small birds. They will occasionally feed on amphibians and reptiles, and sometimes even other snakes. Rattlesnake bites are uncommon (there were three deaths from snake bites reported in all of Canada between 1980 and 2011). In most encounters, the snakes will depart as soon as they can.

Page 166 – Elbow River floods. Between 1883 and 1932, there were seven major flood events. Another major flood occurred in 1950. The last two floods, in 2005, and 2013, caused massive, widespread damage to Calgary and other parts of southern Alberta. The flood in 2005 caused $400 million in damages.

The 2013 event was even more catastrophic. Leading up to June 20 in 2013, huge amounts of rain fell in the Rocky Mountain foothills. In Canmore, located 106 kilometres west of Calgary, over 220 millimetres fell in just 36 hours, nearly half the annual average rainfall. In the town of High River, 68 kilometres south of Calgary, rainfall of 325 millimetres was recorded in less than 48 hours. The rain falling on already saturated ground, coupled with the steep slopes and heavy snowpacks remaining in the Rockies, resulted in a rapid increase in the size and flow of several rivers. On June 21, the Elbow River reached flow rates ten times its average for that time of year. The amount of damage from the flooding in Calgary was over $5,000,000,000.

Page 168 – Cassiar. By the 1970s, Cassiar had a population of around 1,500 and two schools, two churches, a small hospital, a movie house, a swimming pool, a recreation centre, and a hockey rink. The asbestos had the highest quality found anywhere in the world at the time. In the early 1990s, diminished demand for asbestos doomed the community and the mine. The company attempted to boost production by converting from an open-pit operation to an underground mine. The underground mining venture failed miserably. In 1992, Cassiar Asbestos decided to close the mine and liquidate its assets, including the town itself. Most of the contents of the town, including a few houses, were auctioned off and trucked away. Almost all the remaining dwellings were bulldozed and burned to the ground. The mill was briefly reactivated in 1999 by Cassiar Chrysotile Inc., which had a reclamation permit to clean up the site. Eleven thousand tons of asbestos were exported before the mill burned down on Christmas Day of 2000, effectively halting all production. Today the streets are bare, and weeds flourish where the houses once stood. Neglected and in disrepair, the hockey arena collapsed in 2008, and the church met the same fate in 2015. The tramline, which transported ore from the mine down the mountainside to the mill, was purchased in the auction, however, the buyer left it, and it still stands.

Page 169 – jade mining. Jade is either nephrite or jadeite; they have different ranges of colour and form. Nephrite is a variety of the calcium and magnesium-rich amphibole mineral actinolite (aggregates of which also make up one form of asbestos). The other mineral species known as jade is jadeite, which is a variety of pyroxene.

Jadeite and nephrite are part of a group of minerals called amphiboles (the term means double entendre, implying ambiguity). They are characterized by an atomic structure having one atom of silicon surrounded by four oxygen atoms. While nephrite jade possesses mainly grey and green colours (and occasionally yellows, browns, or whites), jadeite jade, which is rarer, can also contain black, red, pink, and violet colours.

Nephrite jade is an ornamental stone used in carvings, beads, and cut gemstones. Nephrite also occurs as a translucent white to very light, yellow form known in China as mutton fat jade. Canada is the principal source of modern lapidary nephrite. Nephrite jade was used mostly in pre-1800 China as well as in New Zealand, the Pacific Coast and Atlantic Coasts of North America, Neolithic Europe, and southeast Asia.

Jade was first identified in Canada by the Chinese settlers in 1886 in British Columbia. At that time, jade was thought to be worthless as miners were focused on finding gold. Jade did not become a commercial stone in Canada until the 1970s. There are over fifty known nephrite deposits in BC. The deposits in southern BC, the Cassiar region, Cry Lake and Dease Lake, and the Mount Ogden areas. These nephrite occurrences consist of individual blocks, talus blocks, boulder fields, and in situ occurrences. Until the 1960s, almost all of the nephrite that was produced in BC came from secondary deposits (material that has been moved by water and deposited in a different location). With the rapid expansion of the amateur lapidary activity after World War II, production in the jade fields of British Columbia's picked up, and they became the most important suppliers. At that time, markets developed in Asia and Germany.

Mining activity for nephrite gradually depleted the secondary deposits, but the increasing value of jade led to exploration in primary deposits (occurring in its source rock, in situ). These efforts uncovered primary deposits adjacent to the Fraser River in southern British Columbia, the Mount Ogden area of central British Columbia, and the Cassiar region.

Today, BC is the main supplier for the China market of nephrite.

Nephrite mining in the mountains is full of risk, uncertainty, and difficult working conditions. Winters are typically long and extremely cold, and deposits of nephrite are remote, so mining can only happen during the summer season, which is only two months. Deposits range in size from 30 cm to four metres wide. Large deposits of jade are challenging to quarry. Nephrite boulders, which are sometimes found above ground, can weigh as much as 180 tonnes and are rarely under 50 kilograms. Typically, the maximum size which is used for transporting boulders by truck is about 5 tonnes. In 2000, an 18-tonne nephrite boulder was found in the Cassiar region and in 2011, in Myanmar, a 192-tonne jadeite boulder, worth an estimated $170 million, was unearthed.

Page 182 – CGL protests. After TC Energy received its partners' go-ahead in November 2018, it appealed to the Supreme Court of British Columbia to grant an injunction to stop the blockade of its intended route through the Wet'suwet'en territory. A temporary injunction was issued in December by BC Supreme Court Judge Marguerite Church to allow CGL pre-construction work.

On January 7, 2019, the RCMP conducted a raid to enforce TC Energy's injunction, removing the barricades on the Morice Forest Service Road and arresting 14 of the Wet'suwet'en land defenders. The RCMP faced criticism from protesters for the amount of force used in the bloodless raid. The RCMP set up a continuous presence along the road, establishing a local detachment called the Community Industry Safety Office. The Wet'suwet'en remained in place along the road but did not make any further efforts to disrupt the CGL pre-construction work.

In December 2019, TC Energy prepared to start construction in the Wet'suwet'en territory. It applied for an extension of the injunction order as the land defenders had resumed blockading access after the pre-construction work was done. This injunction was extended by Judge Church of the BC Supreme Court on

December 31. The extension included an order authorizing the RCMP to enforce the injunction. In her decision, Church stated: "There is a public interest in upholding the rule of law and restraining illegal behaviour and protecting of the right of the public, including the plaintiff, to access on Crown roads," and "the defendants may genuinely believe in their rights under Indigenous law to prevent the plaintiff from entering Dark House territory, but the law does not recognize any right to blockade and obstruct the plaintiff from pursuing lawfully authorized activities." In a public statement, the Wet'suwet'en chief rejected the decision.

On January 1, 2020, after rejecting the injunction, the hereditary chiefs ordered the eviction of the RCMP and Coastal GasLink personnel from the Wet'suwet'en territory. On January 30, the RCMP announced that they would stand down while the hereditary chiefs and the province met to discuss and try to come to an agreement. On February 3, the Office of the Wet'suwet'en asked for a judicial review of the environmental approval for the pipeline. All parties issued statements on February 4, 2020, that the talks had broken down. On February 6, the RCMP began removing the blockades on Wet'suwet'en territory, arresting 28 protesters at camps along the route between February 6 and 9. All were released within two days. The RCMP also detained several reporters and was accused of interfering with the freedom of the press. Union of British Columbia Indian Chiefs Grand Chief Stewart Phillip stated that "we are in absolute outrage and a state of painful anguish as we witness the Wet'suwet'en people having their title and rights brutally trampled on and their right to self-determination denied."

During the enforcement action by the RCMP, a large number of personnel and advanced equipment was used, including heavily armed tactical teams, division liaison personnel, regular uniformed officers, canine units, helicopters, drones and snowmobiles, according to CBC News.

On February 11, the RCMP announced that the road to the

construction site was cleared, and TC Energy announced that work would resume the following Monday. After the hereditary chiefs made it a condition for talks with the government, the RCMP closed their local office and moved all staff resources to their detachment in Houston on February 22. Throughout February and March, solidarity protests and blockades were held across the world. Most in-person actions were halted in mid-March due to the COVID-19 pandemic, but online solidarity rallies continued. On June 5, the BC Prosecution Service issued a statement saying that criminal contempt charges for 22 members of the Wet'suwet'en Nation and their supporters would not be pursued. Additionally, Coastal GasLink issued a statement that they would not pursue civil contempt charges against the protesters.

Protests sprang up across Canada in February, causing massive disruptions to commercial rail transport and public highways. Once the RCMP began to take down the Wet'suwet'en blockades, protests supporting the hereditary chiefs and the land defenders started up. On February 11, protesters surrounded the BC Legislature in Victoria, preventing the traditional ceremonies associated with the reading of the Throne Speech by the Lieutenant Governor. Members of the Legislature had to have police assistance to enter or used alternate entrances. Other protests took place in Kamloops, Chase, Nelson, Calgary, Edmonton, Regina, Winnipeg, Toronto, Ottawa, Hamilton, Montreal, Sherbrooke, and Halifax. Several major protests blocked access to the Port of Vancouver, Deltaport, and two other ports in Metro Vancouver for a number of days before the Metro Vancouver police began enforcing an injunction on the morning of February 10, 2020, arresting 47 protesters who refused to cease obstructing the port.

On February 15, over 200 people protesting in Toronto blocked Macmillan Yard, the second-largest rail classification yard in Canada. On February 16 and 17, protesters temporarily blocked the Rainbow Bridge in Niagara Falls, Ontario and Thousand Islands Bridge in Ivy Lea, Ontario, two major border crossings between the United States and Canada. At the same time, Mi'kmaq demonstrators partially blocked access to the Confederation Bridge, the sole road link to Prince Edward Island. On February 18, several activists were arrested for trespassing at BC Premier Horgan's residence. On February 24, individuals shut down a major highway junction in Hamilton, Ontario. A nationwide student walkout occurred March 4, with university students across the country showing their support for the Wet'suwet'en protesters.

On February 27, 2020, meetings began between the hereditary chiefs and the Canadian and BC governments, represented by Crown-Indigenous Relations Minister Carolyn Bennett and BC Indigenous Relations Minister Scott Fraser. The meetings were planned to last two days. They took place in Smithers, BC. For the duration of those two days, the RCMP agreed to stop all patrols on the Morice West Forest Service Road and to shut down their mobile detachment (CISO) in the area. In addition, Coastal GasLink agreed to suspend operations in the territory during the talks. RCMP and CGL work resumed on the territory once the meetings were complete.

On March 1, after three days of meetings, the Canadian Crown-Indigenous Relations Minister, the BC Indigenous Relations Minister and representatives of the Wet'suwet'en, including hereditary chiefs and matriarchs announced a proposed agreement to address the Wet'suwet'en land rights, title, and a protocol for addressing any future projects impacting on their territory, in the form of an MOU (memorandum of understanding). Specific details of the agreement were not immediately released because the MOU had to first be seen and ratified by the broader Wet'suwet'en nation. However, it was made clear by all parties to the discussions that the agreement did not concern the Coastal GasLink Pipeline project. On March 10, a statement was released by Theresa Tait-Day, president and one of three founders of the Wet'suwet'en Matrilineal Coalition, who was stripped of the subchief name Wi'hali'yte in the mid-2010s over her support of the pipeline project and suspected conflict of interest. Tait-Day stated that the proposed agreement made on March 1 was not inclusive of the entire community, saying, "the government has legitimized the meeting with the five hereditary chiefs and left out their entire community. We can not be dictated to by a group of five guys." According to Tait-Day, "over 80 percent of the people in our community said they wanted LNG [First Nations LNG Alliance] to proceed."

Meetings were held among the individual clans of the Wet'suwet'en throughout March. The memorandum of understanding was ratified by the attendees of one meeting of the Laksilyu (Small Frog Clan). They held a meeting via Zoom on April 22. According to the hereditary chiefs, the Gilseyhu (Big Frog Clan) met once and endorsed the MOU, as did the Laksamshu (Fireweed and Owl Clan) and the Tsayu (Beaver Clan) after two meetings.[41] The Gitdumden (Wolf and Bear Clan) met twice, but their third meeting (scheduled for March 16) was cancelled due to a death in the community.

On April 30, the hereditary chiefs made a joint statement with the provincial and federal governments that all five clans had agreed to ratify the MOU after having reviewed and discussed it in the prior weeks. However, the elected chiefs of five Wet'suwet'en band governments (Nee Tahi Buhn Indian Band, Skin Tyee Nation, Ts'il Kaz Koh First Nation, Wet'suwet'en First Nation, and Witset First Nation) released their own joint statement in response the following day, calling on the agreement to be withdrawn, saying they weren't consulted properly. A further statement released on May 11 called once again for the agreement to be withdrawn so the elected governments could be consulted properly, and further calling for Minister Bennett to resign. The May 11 statement was not signed by Chief Sandra George of Witset or Chief Cynthia Joseph of Hagwilget. The draft agreement was finally distributed to the elected band councils on May 7, to all other Wet'suwet'en the following day, and finally, it was published on the Office of the Wet'suwet'en website on May 12. The MOU was signed by hereditary chiefs Minister Bennett, and Minister Fraser on May 14, 2020, in a virtual ceremony via Zoom. The memorandum did not address the CGL pipeline project, nor did it alter Wet'suwet'en rights and title. The MOU stated that the Canadian and British Columbian governments recognize that those rights and title are held under the Wet'suwet'en's own system of governance, and commits Canada and BC to a three-month process to craft a formal Affirmation Agreement that confirm aboriginal title as a legal right. It also established a twelve-month timeline for negotiation on jurisdiction including over land-use planning, resources, water, wildlife, fish, and child and family wellness. Further, it acknowledged that reunification of the rift between the hereditary leadership and the elected band councils is an essential part of the implementation of the MOU.

Page 194 – The title of the report was Shrub Monitoring in Canada's Arctic Using Multi-Scale Measurements from Field Plots, Unmanned Aerial Vehicles and Satellite Remote Sensing (Clement and Fraser 2017). "These combined plot surveys begin to build a baseline from which vegetation changes can be monitored within the Cambridge Bay area. They also demonstrate a potential strategy for combining conventional in situ vegetation monitoring with new low-cost UAV mapping methods, which could be applied to other sites within Canada's Arctic. The close proximity of the vegetation plots to Cambridge Bay promotes their use in educational and outreach activities associated with the new Canadian High Arctic Research Station (CHARS).

Recent climate warming has been pronounced in the Arctic, with the average annual temperature increasing between 2-3°C since the 1950s (Arctic Climate Impact Assessment [ACIA] 2004). Accelerated temperature increases have driven declines in Arctic sea ice extent, snow cover duration, glacier mass, and increases in permafrost temperatures (ACIA 2004). The terrestrial Arctic has also undergone important temperature-driven changes in the form of increased thermokarst, more intense tundra fires, and enhanced vegetation productivity. Strong evidence for an Arctic-wide expansion of upright shrubs (called shrubification) comes from a variety of sources, including analysis of satellite imagery, historical photographs, vegetation plots, and growth rings (Myers-Smith et al. 2011). In order to assess shrubification, the project is addressing the productive "subzones" defined in the Circumpolar Arctic Vegetation Map (CAVM Team 2003).

Shrub changes will have a range of implications for climate feedbacks, infrastructure, and wildlife. Arctic shrubs can respond rapidly to favourable growing conditions and therefore provide a sensitive indicator of climate-driven vegetation change. Shrub growth is also stimulated locally by a variety of natural and human-caused disturbances, including thaw slumps, wildfires, gas well sumps, and other infrastructure that may result in increased snow accumulation and ground warming. Expanding shrubs also overtop and out-compete terrestrial lichens, potentially reducing the quantity of this high-energy forage for barren-ground caribou.

Shrub changes can be measured over a range of spatial scales: at local scales using field plot measurements; over landscapes using repeat aerial and ground photography; and at regional-to-continental scales based on analysis of multi-temporal satellite data (Myers-Smith et al. 2011, Henry et al. 2012). Optical satellite imagery is particularly sensitive to tundra vegetation leaf area and above-ground biomass, making shrub cover amenable to mapping and repeat measurement over large areas (Riedel et al. 2005).

Our ability to document Arctic shrub and other vegetation changes is currently limited in Canada. At the finest scale of measurement (e.g. vegetation plots), there are a small number of locations where changes in tundra shrubs are being experimentally measured. At coarser spatial scales, there have been a limited number of remote sensing studies documenting regional vegetation greening that in many cases is likely related to shrub proliferation (i.e. Fraser et al. 2014, Ju and Masek 2016). Ecological modelling predicts large-scale northward shifts in Arctic vegetation based on climate change scenarios. Yet, there is limited understanding of which regions in Canada are likely to show the greatest increases in shrub growth.

Community-based monitoring guided by traditional knowledge holders represents a powerful means of documenting environmental change at select sites across Canada's Arctic. Observations by Canada's Inuit have provided some of the strongest evidence for climate-driven vegetation changes. To facilitate community-based monitoring that involves northerners with insights regarding environmental change, the University of Victoria (UVic) recently established a plot-based vegetation and permafrost monitoring protocol (Northwest Territories – Cumulative Impact Monitoring Program [NWT-CIMP]) (Lantz et al. 2013 – included as a supplementary file to this report) that can be carried out by a range of participants (community members, university and government scientists, government inspectors, etc.). Between 2010 and 2014, the NWT-CIMP program established 53 environmental monitoring sites in the Mackenzie Delta region to characterize baseline variation in vegetation, snow, and permafrost (the approach and work by Lantz et al. is documented in a separate report).

The 20 x 20 metre vegetation plots were situated near the road about ten kilometres northeast of Cambridge Bay so that they could be easily revisited. They were selected to represent common vegetation communities in the Cambridge Bay region and to be sufficiently distant from Cambridge Bay so that Unmanned Aerial Vehicle (UAV) surveys could be conducted under a Transport Canada, Special Flight Operations Certificate (SFOC) exemption.

The vegetation plots were surveyed using both conventional, in situ (on site or in place) methods and new UAV-based photogrammetric mapping methods. The in situ methods followed a published monitoring protocol (Lantz et al. 2013) and involved measuring percent vegetation cover by both plant functional groups and more detailed species community composition. The UAV photogrammetric surveys were used to generate sub-centimetre resolution orthomosaics and 1 centimetre resolution vegetation height models. The vegetation height model was compared to a sample of 20 willow dwarf shrub heights measured within one of the plots.

The beginning point for each plot was the northwest (NW) corner which was selected on the basis of homogeneity of landscape conditions (slope, aspect, ground-surface conditions, and distribution of vegetation). Working in a counter clock-wise direction, the four sides (20 metres [m] each) of the plot were located using a Silva compass and 30 m tape running in cardinal directions (east, south, west, and north). Each plot was 20 m x 20 m, which resulted in the creation of four subplots (SP1 to SP4). At each corner, a 1 m long round iron bar was pounded into the ground using a small sledgehammer and a ¾ inch piece of polyvinyl chloride pipe (PVC) was fitted over the bar. The top 50 centimetres (cm) of bar and PVC was left above ground and marked with double pink flagging tape. A thin metal stake (approximately 40 cm x 2 millimetres [mm]), with pink flagging attached, was inserted at ten m intervals on each side and in the middle of the plot to show the extent of each SP.

Functional group frequency provides an estimate of generalized vegetation cover, measured inside the 20 m x 20 m plot. The functional groups, which included tall shrubs (> 40 cm), dwarf shrubs (< 40 cm) Substrate types, including soil, rock, and litter, were also described in the functional group assessment. Within each SP, functional groups and substrate were assessed at 16 points (shown in Figure 2). A thin metal stake was lowered at the sample point, and functional groups and substrate were recorded.

Community composition data was collected at four sample points (at the 10 m point along each side of the sample plot) using a 1 m x 1 m frame. It should be noted that the community composition data collected by Lantz et al. employed a 1.1 m x 1.1 m frame. The 1.1 m square allows for creation of a 100-point grid using nylon line tied across the frame at 10 cm intervals. The 1 m frame, which was built by PKC staff, creates 81 sample points. Ultimately, the 81 sample point data was all converted to frequency ratings (percentage of occurrence)."

Page 194 – Inuit carrot. [information provided here is taken from *Flora of the Canadian Arctic Archipelago* (Aiken et al. 2011)]. "The part of the root between the bottom of the stem and the root (caudex) is called tuqtaw. The roots can be quite large. They can be eaten raw and have a sweet taste. The roots are yellowish or greenish when they are young. When they are older, they are brownish and more fibrous. They are good for stomach aches and are used to control diarrhoea (Ootoova et al. 2001). The roots are excellent fried in seal oil or store-bought oils (Eva Aariak, personal communication, 2006). Airait were sucked by babies being carried in amautiit. Because the babies like the itsii (juice), it would help them fall asleep. The yellow flowers are eaten by geese. When there is blubber available, the roots of this plant are used as food. You would cook the blubber and fry the roots, which made them soft. Eating them straight might not be good, but when mixed with blubber, they make good eating (Z. Innuksuk, personal communication, reported in Mallory and Aiken 2004). We ate the roots fried in seal fat as soon as the snow melted (Tuqassie Killiktee, personal communication, reported in Mallory and Aiken 2004)."

Page 203 – Construction of RCAF Station Cold Lake, located southwest of the range in Alberta, began in 1952. It became a major Cold War facility after 1954, eventually housing more than 2,000 service personnel. In operational terms, evaluation facilities were constructed on the accompanying range so that fighter pilots could test armament systems and fire their guns at various altitudes. Non-military use of the CLAWR has increased over the last decade and continues to grow as different resource sectors vie for access to land

and resources (such as natural gas, commercial fishing, and logging) in and around the range.

Page 210 – In 2004, there were 4,100 people admitted to hospitals in Canada from ATV incidents. Between 2007 and 2011, there were 912 fatalities. Those statistics are staggering. In the United States, the numbers are even more alarming. Statistics released by the Consumer Product Safety Commission showed that in 2005, there were an estimated 136,700 injuries associated with ATVs treated in US hospital emergency rooms. In 2004, the latest year for which estimates were available, 767 people died in ATV-associated accidents.

Page 212 – Canola and rapeseed. The name for rapeseed comes from the Latin word rapum meaning turnip. Turnip, rutabaga (swede), cabbage, Brussels sprouts, and mustard are related to rapeseed and canola. Rapeseed also belongs to the Brassica genus. Brassica oilseed varieties are some of the oldest plants cultivated by humanity, with documentation of their use in India 4,000 years ago, and in China and Japan 2000 years ago. Its use in Northern Europe for oil lamps is documented from the 13th century.

Rapeseed oil extracts were first put on the market in 1956–1957 as food products, but these suffered from several unacceptable characteristics. Rapeseed oil had a distinctive taste and a disagreeable greenish colour, due to the presence of chlorophyll. It also contained a high concentration of erucic acid. Canola was bred from rapeseed cultivars of *B. napus* and *B. rapa* at the University of Manitoba in the 1970s.

Canola was originally a trademark name of the Rapeseed Association of Canada, and the name was a condensation of 'Can' from Canada and 'OLA' meaning 'Oil, low acid', but is now a generic term for edible varieties of rapeseed oil in North America and Australasia. The change in name serves to distinguish it from natural rapeseed oil, which has a much higher erucic acid content. A genetically engineered rapeseed that is tolerant to herbicide was first introduced to Canada in 1995 (Roundup Ready canola). A genetically modified variety developed in 1998 is supposed to be the most disease- and drought-resistant canola variety to date. In 2009, 90% of the Canadian crop was herbicide-tolerant. In 2005, 87% of the canola grown in the US was genetically modified. In 2011, out of the 31 million hectares of canola grown worldwide, 8.2 million (26%) were genetically modified. In 2014, world production of rapeseed oil was 26 million tonnes, led by China, Germany, and Canada as leading producers accounting together for 47% of the world total. Canada was the world's largest exporter of rapeseed oil in 2016, exporting 2.9 million tonnes or approximately 94% of its total production. In China, rapeseed meal is mostly used as a soil fertilizer rather than for animal feed, while canola is used mainly for frying food.

Page 217 – On June 18, 2013, Kinder Morgan, the original owner of the pipeline, filed an application with the Canadian National Energy Board pursuant to Part III of the National Energy Board Act to build a second pipeline under the title of the TM Pipeline Expansion project. The second pipeline was to run parallel to the existing pipeline, between Edmonton and Burnaby, (east of Vancouver) and to be used to transport diluted bitumen. The additional pipeline required 12 new pumping stations. The proposed expansion, with 980 kilometres of pipe, would increase the system's capacity from 300,000 barrels per day to 890,000 barrels per day. An investment of $7.4 billion would be required to complete the twinning project.

Kinder Morgan had the support at that time of numerous large petroleum industry customers for this expansion (BP Canada Energy Trading Co., Canadian Natural Resources Ltd., Canadian Oil Sands Ltd., Cenovus Energy Inc., Devon Canada Corp., Husky Energy Marketing Inc., Imperial Oil Ltd., Nexen Marketing Inc., Statoil Canada Ltd., Suncor Energy Marketing Inc., Suncor Energy Products Partnership, Tesoro Refining & Marketing Co, and Total E&P Canada Ltd).

In 2016, the BC government stated it did not support the TM project, partly because Kinder Morgan has not provided enough information about its proposed spill prevention and spill clean-up program. On November 29, 2016, Canada's federal cabinet approved the expansion project, announcing that the approval was "subject to 157 binding conditions that will address potential Indigenous, socio-economic and environmental impacts, including project engineering, safety and emergency preparedness."

On January 11, 2017, B.C. Premier Christy Clark announced British Columbia's support for the expansion of the pipeline, saying the project met her government's five conditions for approval and included a revenue-sharing agreement worth up to $1 billion.

Kinder Morgan bailed on the project after shareholders voted overwhelmingly to approve the sale of the TM pipeline and expansion project to the Canadian government for $4.5 billion. The vote came just 30 minutes after the Federal Court of Appeal quashed the federal government's approvals to build the massive project, handing a huge victory to Indigenous groups and environmentalists opposed to it. In May 2018, the federal government purchased TM from the Houston-based Kinder Morgan for $4.5 billion. At that time, the government said it would seek outside investors to complete the expansion. These investors would also be indemnified for any delays induced by provincial or municipal governments.

In 2020, three insurance companies that had previously supported the project withdrew their support, including Zurich Insurance Group, the leading insurer.

In a September 2020 interview with the Canadian Press, TM's CEO, Ian Anderson, said that the expansion was on schedule despite a "$5.2-billion rise in its estimated cost to $12.6 billion." Anderson cited other challenges to construction, including the COVID-19 pandemic, the slump in demand for fuel which contributes to the slump in the price of oil, and the ongoing protests by opponents to the expansion. Protests have been ongoing periodically since 2012.

Chronology of Projects

1970 BC Ministry of Agriculture – Keremeos Soil Survey and First Field Lessons

1971 BC Ministry of Agriculture – West Kootenays Forestry Plots

1972 BC Ministry of Agriculture – Nelson Map Sheet, Ecosystem Plots

1973 BC Ministry of Environment – Shuswap Ecosystem Sampling

1975 BC Ministry of Environment – East Kootenays Biophysical

1976 BC Ministry of Environment – Northeast Coal Project Biophysical Inventory

1977 BC Ministry of Environment – Northeast Coal Extension

1978 BC Ministry of Environment – Galiano Island Resource Inventory and Level
Mountain/Horseranch Range Biophysical

1981 Pedology Consultants – Canstar Biophysical

1982 Pedology Consultants – Stikine River Drainage Broad Biophysical and Tahshish-Kois Ecosystem Sampling

1983 Surviving the Recession

1984 Shearwater Ecological Services – Kimsquit River Ecosystem Sampling

1985 Shearwater Ecological Services – Bell-Irving Wildlife Habitat Mapping and Observatory Inlet Biogeoclimatic Sampling

1987 Shearwater Ecological Services – Impacts of Glyphosate on Grizzly Bear Habitat

1988 Shearwater Ecological Services – Itchas and Ilgachuz

Caribou Habitat Mapping and Ecosystems of TFL 41

1989 Shearwater Ecological Services – Khuzeymateen River Ecosystem Mapping

1990 Shearwater Mapping Ltd. – Gwai Hannas Biophysical Mapping

1991 Shearwater Mapping Ltd. – Iskut River and Lower Stikine Biophysical

1992 Shearwater Mapping Ltd. – Northern Rockies Wildlife Habitat Mapping, Gwai Haanas Biophysical and McGregor Model Forest Biophysical

1993 Shearwater Mapping Ltd. – Tweedsmuir-Atnarko Grizzly Habitat Mapping

1994 Shearwater Mapping Ltd. – Dog Creek Biophysical Project

1995 Shearwater Mapping Ltd. – Start of Terrestrial Ecosystem Mapping Program, Sensitive Ecosystem Inventory in the Capital Region

1996 Shearwater Mapping Ltd. – Ongoing Work for MacBlo and the Itcha-Ilagachuz Biophysical

1997 Shearwater Mapping Ltd. – Ongoing MacBlo Projects, Itcha-Ilgachuz Ranges Biophysical, Snake-Sahteneh Biophysial

1998 Shearwater Mapping Ltd. – Ongoing MacBlo Projects and Charlotte Alplands Biophysical

1999 Shearwater Mapping Ltd. – Ongoing MacBlo Projects including Port McNeil and the Sarita River Drainage

2001 Shearwater Mapping Ltd. – Ecosystem Mapping of a Portion of Savary Island and TFL 33 TEM

2002 Shearwater Mapping Ltd. – Merritt TSA PEM Auditing

2004 Clement Consulting – Queen Charlotte TSA TEM for Husby Forest Products, Graham Island

2005 Clement Consulting – Training Natural Resource

Technicians for the Upper Nicola Band

2007 Clement Consulting – Harper Creek TEM for Dillon Consulting Ltd.

2008 to 2009 Clement Consulting and Dillion Consulting Ltd. – Looking for a Full-Time Job

2009 to 2010 Dillon Consulting Ltd. – Small Environmental Projects in Calgary

2010 Dillon Consulting Ltd. – Biophysical Survey near Cassiar for a Proposed Mine

2011 to 2013 EBA – Transportation Regulatory Approvals

2014 Stantec Consulting Ltd. – Prince Rupert Gas Terminal and Uplands Pipeline

2016 Clement Consulting – Quadra Island Carbon Offsetting

2016 Chris Clement Eco Solutions Ltd. – Small Contract Jobs and Research in the High Arctic

2017 Chris Clement Eco Solutions Ltd. – Small Contract Jobs, DHD and Canadian Tire

2018 Chris Clement Eco Solutions Ltd. – Small Contract Jobs, Canadian Tire

2019 Chris Clement Eco Solutions Ltd. – Small Contract Jobs, Canadian Tire

2020 Black Fly Environmental – Trans Mountain, High Elevation Grassland TEM, BC Gov. Caribou Arboreal Lichen Habitat

2021 Black Fly Environmental
Continuation of High Elevation Grassland TEM

Reviews

Chris Clement, who I have known for over fifty years, is an excellent field ecologist. Chris is very knowledgeable about ecosystems throughout British Columbia and well beyond. His enthusiasm to understand all aspects of ecosystems has made him an effective teacher and mentor for a new generation of ecologists. Mr. Clement's book reveals his exciting adventures throughout his years of field ecology, along with the interesting individuals who made this journey with him.

– Ted Lea, Terrestrial Ecologist, Victoria, BC

With this epic work Chris captures 51 years of surveying and studying western Canada's wilderness in a very engaging, provocative, and sometimes humorous manner. The range and uniqueness of its ecosystems provides a backdrop to each adventure. Historical anecdotes add interesting context. A professional but personal view on resource development and related projects are raised to be considered. What is clear is his love of wilderness, the profession that makes it his office and the colleagues with whom he shares both.

– Evert Kenk, retired BC Government Executive Director and co-editor of the Terrain Classification System for British Columbia

Reviews

I was lucky to work with Chris in his 51st field year, part of which was spent scampering through the rugged subalpine and alpine grasslands of the Elk Valley. Evenings after the work day, Chris could be found mentoring junior staff on plant identification or regaling them with stories of adventure from his rich career. His youthful love of fieldwork is infectious. This book is a treasury of field wisdom and a testament to a career well spent outdoors.

– Iraleigh Anderson, Plant Ecologist

Printed in Great Britain
by Amazon

43275136R00175